Guidelines for

PREPARING PROPOSALS

Second Edition

Roy Meador

 LEWIS PUBLISHERS

Library of Congress Cataloging-in-Publication Data

Meador, Roy
    Guidelines for preparing proposals / Roy Meador. — 2nd ed.
        p. cm.
    Includes bibliographical references and index.
    1. Proposal writing in business.   2. Proposal writing in
research.
    I. Title.
    HF5718.5.M43   1991
    658.15′224 — dc20   91–24732
    ISBN 0-87371-588-8

LEWIS PUBLISHERS, INC.
121 South Main Street, Chelsea, Michigan 48118

Printed in the United States of America   2 3 4 5 6 7 8 9 0

Dedicated to

Edward E. Lewis,
Science Editor
Friend
Wise and Jovial Companion in the
Great Journey of Words, Work, Books, and Ideas

Roy Meador

Roy Meador is a freelance writer who specializes in technical and scientific subjects. He has written extensively for government and corporate clients including Gelman Sciences, General Motors, MERRA, Pfizer Inc., the U.S. Department of Energy, the U.S. Small Business Administration, NASA, the State of Michigan, the Michigan Small Business Development Center Network, the Engineering Society, and many others.

His experience includes preparation of numerous proposals for universities, companies, and associations. He served as a writer on proposal preparation teams for a wide variety of topics including acid rain research, appropriate technology, and computer-based process control systems. He assisted in developing applications responding to public and private grant programs, among them the Small Business Innovation Research (SBIR) funding opportunity sponsored by the federal government.

Mr. Meador's credits include completing a writing assignment for the National Aeronautics and Space Administration (NASA) on the utilization of LANDSAT, earth-scanning satellites, and a report on district heating for the Argonne National Laboratory and the U.S. Department of Energy. He researched and wrote publications on gasohol, low-level radioactive waste disposal, small business research and development (R&D), innovation, conventional defense activities, dual-use technology, R&D opportunities for businesses, and starting a business.

The author is a graduate of the University of Southern California in Los Angeles and Columbia University in New York City. He served as a line officer and navigator in the U.S. Navy during the Korean War. In addition to proposals, he has produced technical and annual reports, manuals, brochures, newsletters, articles, speeches, and related forms of professional writing in carrying out public and private sector writing assignments.

He has published books on alternate energy, the implications of modern technology, Benjamin Franklin's science, and the widely used first edition of *Guidelines for Preparing Proposals*. His articles have appeared in many national periodicals, among them *The New York Times, Chicago Tribune, Smithsonian, The Christian Science Monitor, Chess Life, Analog Yearbook,* and *Newsday.* Complementary interests to his writing include science and technology, book collecting, chess, music, literature, and history.

CONTENTS

CHAPTER 2

DIMENSIONS OF THE NEED AND OPPORTUNITY

"Often the researcher and his director have to seek funds from grant-giving bodies, many of which dispense funds of government origin."
W.I.B. Beveridge
Seeds of Discovery

"The National Science Foundation (NSF) funds research in all fields of science and engineering. . . NSF welcomes proposals on behalf of all qualified scientists and engineers."
National Science Foundation
Grants for Scientific and Engineering Research

A 1931 song from the revue *Crazy Quilt* with words and music by Mort Dixon, Billy Rose, and Harry Warren celebrated the finding of a "Million-Dollar Baby in a Five and Ten Cent Store." During the 1940s, the Andrews Sisters popularized "The Money Song;" and Richard Rodgers and Lorenz Hart once produced a hit tune in about an hour that they called "Ten Cents a Dance."

Thus music and money, from a dime to a million, came together in song. Science and money also make beautiful music together, but the lyrics have to be written in the form of proposals. A million-dollar baby isn't all that remarkable now. A multimillion dollar return is more in order these days through government, foundation, institutional, and industrial grants.

Item:

The 1991 edition of *The Foundation Directory* gave details on 7,581 foundations, including over 900 corporate foundations (each with assets exceeding $1 million or annual grants exceeding $100,000). Foundations in the *Directory* together hold almost $125 billion in assets and award nearly $7 billion in grants each year. Successful proposals win these grants.

Item:

Research Institutes such as the Electric Power Research Institute (EPRI), supported by U.S. electric utilities, and the Gas Research Institute (GRI), maintained by the gas industry, invest large sums each year in research. Suitable projects receive institute funding as a result of effective proposals.

Item:

Modern industry is a major customer and vast market for R&D, services, and products from both individuals and companies. One prime example is represented by the subcontracting opportunities for small businesses with U.S. Department of Defense (DOD) Prime Contractors. The objective, according to the DOD *Small Business Subcontracting Directory*, is "to expand the industrial base by involving a greater share of this nation's small business community in providing the goods and services required for national security."

In FY 1989, according to the Directory, DOD Prime Contractors awarded $56 billion in subcontracts of which $22 billion went to small businesses. DOD also emphasized the option of direct prime contracting opportunities for small businesses which received awards of $22.8 billion in prime contracts during FY 1989.

Information about such opportunities are available from Small Business Specialists at U.S. military installations. Copies of the periodically published DOD *Small Business Subcontracting Directory* are available from The Superintendent of Documents, U.S. Government Printing Office, Washington, D.C. 20402.

Good proposals are the primary means of initiating action on these prime contracting and subcontracting opportunities. Proposals are also the vehicles needed by individuals and companies to approach large corporations and effectively make known what they have to offer that large organizations need.

Item:

In 1982, the U.S. Congress enacted the Small Business Innovation Development Act to stimulate innovation and technological development through small businesses. Under this legislation, each federal department or agency that disburses at least $100 million a year in R&D awards and contracts must set aside R&D award money specifically earmarked for qualified small businesses, university researchers, and individual entrepreneurs.

Through this legislation (Public Law 97–219), the country benefits from a vigorous stream of new scientific and technological developments. Small businesses receive welcome injections of badly needed R&D funds. The program grew annually, and by 1987, the funds that federal departments and agencies were required to set aside for small business R&D awards reached nearly half a billion dollars. This small business R&D money is allocated to those who submit sound and convincing Small Business Innovation Research (SBIR) proposals.

In 1986, Congress by a large majority reauthorized and the President signed an extension of the SBIR Program until the end of FY 1993. SBIR proposals during the

1987–1993 cycle compete for awards from a total SBIR money mountain that is expected to exceed $3 billion. Early in the 1990s, concerted efforts were progressing to extend the SBIR Program beyond FY 1993.

In *Scientific American*, January 1991, B. R. Inman, President of Inman Associates, and Daniel F. Burton, Executive Vice President of the Council on Competitiveness, warned that international competition in many key areas—semiconductors, super-conductors, electronics, high-definition television—is effectively firing a challenging shot across American industry's bow. Inman and Burton stressed the need for widening "R&D efforts to include more commercially relevant technology." They wrote, "Government agencies will have to be refocused to address the new priority of technology and competitiveness." The SBIR Program represents a small but strong step in this area.

Paul Wallich and Elizabeth Corcoran in the same issue of *Scientific American* reported, "High-technology manufacturing by small companies is growing significantly faster than is manufacturing as a whole." High-technology R&D by small companies is also increasingly productive due to the SBIR Program and complementary public and private R&D funding sources. R&D funds from these sources are mainly distributed on the basis of proposals.

R&D Grants Keep Small Businesses Growing

"The business of government is to keep the government out of business—that is, unless business needs government aid," Will Rogers mused early in the 20th century. Years later Norman Cousins pointed out that, like it or lump it, the U. S. government is a partner in practically every American business. State governments also frequently declare themselves voting partners as well. It can be argued that this isn't a bad thing when federal and state governments come through with funds for long-term support of innovative science and technological progress more generously than private industry, which may be caught up in the anxious pursuit of immediate profits.

The U. S. government and many state governments are a large and steady source of R&D funds. To participate, you need not wait to receive a formal "Request for Proposal" or RFP from a government agency. Most government agencies encourage submission of "unsolicited proposals" if you have something appropriate to recommend that contributes to the fulfillment of a government mission.

Thousands of small businesses started and stay alive thanks to their successes with solicited and unsolicited proposals. The SBIR Program has been particularly beneficial and supportive in this connection. The following statements are representative tributes by high-technology business founders and entrepreneurs:

"Our goal is to sell products resulting from the government R&D support that has kept us going. SBIR and the State Research Fund (SRF) provided a bridge from the university to the business world. We spent a lot of the first year writing grant proposals. We learned from rejections and won our first SRF grant and first SBIR Phase I award with 1986 proposals."

Marilyn Katz-Pek
Cofounder of BioQuant, Inc.

Marilyn Katz-Pek indicated that the company's first successful proposals allowed the company to open its initial laboratory and hire its first employee.

"Without the SBIR and SRF Programs, SoloHill would not exist. R&D got us where we are today. We are selling our microcarrier beads for biotechnology applications throughout the world. Our success in qualifying for government R&D funds helped our company survive in the beginning and emerge from the normal round of early mistakes. We learned if you develop a history of winning, you find it gets easier to win. Even if you start out making mistakes in preparing proposals, you'll get better each time. Keep trying. The results are worthwhile."

David E. Solomon
President, SoloHill Engineering, Inc.

"The government R&D programs allow us to have R&D projects in new areas that we might not enter without the government support provided. What we learn in the course of doing a project may be just as significant as the project itself. These can be secondary and tertiary benefits that keep paying dividends."

Michael E. Korybalski
President, Mechanical Dynamics, Inc.

"I believe strongly in the whole concept of SBIR and SRF. They help a small business like TE Technology take something from the concept stage to the marketable product. They provide just enough to go for it. Thanks to our SBIR experience, we now see that we have the capacity, capability, and technology to serve big companies."

Richard J. Buist
President, TE Technology, Inc.

Richard Buist commented that success in the SBIR and his state's SRF programs gave his company credibility that assisted in obtaining investment capital. He described the first award as "a little seed money going a remarkably long way."

Ed Zimmer, Editor of *The Inventor-Entrepreneur Network Newsletter*, reported in the December 1990 issue, "TE Technology's interest in, and strategy using, the SBIR Program is obvious. The SBIR solicitations are the government agencies' surface-level 'want-list.' The company simply scans these solicitations, sees where their technology may apply, and knocks out a proposal. If the proposal 'wins,' they have the funds to develop the feasibility of a new idea enough to see whether they can sell it. If the proposal 'loses,' they've still made contacts in the agency, established themselves as a credible resource in the technology, gained better insights into the agency's needs, and have educated at least some of the agency's people to the potential benefits of the technology. They win if the proposal wins; they win if the proposal loses."

Richard Buist also emphasized that winning isn't everything. The proposal-writing exercise itself often proves useful and valuable. "Whether it succeeds or not, just preparing a proposal is good for you. You have to crystallize your ideas, devise an action plan, and consider commercialization," he stated.

> *"A phrase that serves us well is 'Stick to your knitting.' This means sticking to what you know and know well. We feel this is an important factor behind our high success rate in the SBIR Program."*
>
> > Dr. James Sheerin
> > KMS Fusion, Inc.

> *"First of all we don't shotgun. We use a rifle with a telescopic sight. We go only after those projects that we see having some benefit for us five years from now."*
>
> > Thomas Ory
> > Daedalus Enterprises, Inc.

> *"In the SBIR Program, if they ask us to do five things, we do six. That's become our standard. If Phase I requires a paper project, we build software. This costs us but the results are worth it."*
>
> > Dr. Omar Helferich
> > Dialog Systems/A. T. Kearney

Listen to these winners. They can help make you a winner with your projects and proposals. But when the subject is winning, Mark Twain's 1896 advice never goes out of style: "There's many a way to win in this world, but none of them is worth much, without good hard work back of it." The generalization is undoubtedly true that back of every successful proposal—past, present, future—is plenty of hard work.

Proposals That Work

The secret of winning government grants is no secret. Effective proposals do the job by ringing the right bells, pushing the correct buttons, and opening key doors that lead to the money. At a series of High Tech Conferences, the following comments were made about the government proposal process. These comments provide substance for reflection—and action—by those already in or planning to enter the annual grant pursuit. In this arena, the race is typically not to the fastest, the strongest, or the luckiest. The race is to the smartest with the best proposals.

Mark H. Clevey, Director of the MERRA Specialty Business Development Center, said at the Michigan High Tech 90 Conference, "While

R&D is important for most industries, it is essential for small emerging firms because their 'lifeblood' is new proprietary technological innovations resulting from R&D activities. Because private sector funding for R&D in the U.S. is limited, public sector R&D grants are a competitive necessity for proprietary technology-based firms."

"The competition is tough in the SBIR Program, but where else can you get half a million dollars to pursue a bright idea? Some companies win more than one SBIR, and the program gives them substantial advanced research budgets. SBIR is cutting edge research. We're after innovative, high risk, high payoff projects with commercial potential."

Roland Tibbetts
National Science Foundation

"The SBIR Program has set off a sharp competition nationwide. We're looking for the best research and think we're finding it. Your proposal to the National Science Foundation should have the kind of excellence you must put into a paper for a refereed journal."

Ritchie Coryell
National Science Foundation

"The main rule — follow instructions and use your common sense."

David Van Meter
U.S. Department of Transportation

"Reauthorization of the SBIR Program passed 421 to 1 in the House of Representatives and unanimously in the Senate. Congress expects the program to produce excellent results. We think it is doing that. Of the proposals we received in the 1982 SBIR Program, about one-third reinvented the wheel in the sense of repeating something already done. Check what's been done. Keep up-to-date. And prepare proposals that convince us you have ideas the Department needs."

Horace J. Crouch
U.S. Department of Defense

"I'm not here to help you. I'm here to get your help in the form of innovative R&D to carry out DOD's mission. The key to success is to learn all you can about Defense needs. At DOD the Work Plan is the most important part of your SBIR proposal and should represent your main effort. What we look for in your proposal is technical credibility."

Robert B. Wrenn
U.S. Department of Defense

"To save time and effort, get in touch with us before writing your proposal and learn what fits our needs at the moment. If you find that you're not under the umbrella of the SBIR because of the specific topics, you have the option of trying the unsolicited proposal route. We look forward to hearing from you."

Harris Coleman
U.S. Nuclear Regulatory Commission

"The process is made more difficult when some small businesses don't pay close attention to what we ask. Please invest the necessary effort in preparing your proposals. We'll pay the price for a first-class proposal. Quality determines what we pay."

Earl Langenbeck
Department of the Navy

"We emphasize quality research. We can't fund junk."

Eugene Steadman
Department of the Air Force

"In preparing your proposal, identify very clearly what your idea is. Provide a clear explanation that will persuade a very knowledgeable person your idea will work, that you have an achievable goal. . .What's an innovation? You tell us. If you get an idea, try it on others, seek critiques, then follow instructions and send us your proposal. But remember, just because it's an innovation, it's not necessarily better."

John Del Gobbo
NASA

"SBIR is not for those with half-baked ideas. This program is for the best in the business. One-third of corporate products during the next decade will come from new products. This is a time of rapid change and turmoil. Product life cycles used to be measured in decades. Now the cycle is much shorter. The U.S. must learn to do business in ways it never did before. SBIR is about new products and processes. That's why large companies are very interested in SBIR winners."

Ann Eskesen
President, Innovation Development
Institute

Ann Eskesen stressed that a successful SBIR proposal must offer clear objectives to the government customer and reflect full awareness and understanding of the government agency's needs and wants. Talk their language to be heard, she advised. "If they are used to doing business with chickens, and you're a duck, you should learn to talk chicken," Eskesen suggested. She called the actual writing of the proposal the end of an extended project-planning process. "Your proposal just puts in order all the planning you did in advance," she said.

Todd Anuskiewicz, Executive Vice President at MERRA, a consortium of government, business, and university representatives, is an authority on the preparation and review of grant proposals. He summarized the endeavor as follows:

"The key to winning grants is having relevant research, completing proposals in accordance with instructions, and submitting them on time. To succeed, learn the exact needs of those providing R&D funds, organize your proposals accordingly, and follow directions precisely. . .It generally pays to start working on your proposals as much ahead of time as possible."

Todd Anuskiewicz
Executive Vice President, MERRA

Anuskiewicz indicated that months typically are required to plan research projects, organize the necessary materials, and conscientiously carry out the drafting, reviewing, and finalizing steps that lead to the completion of a proposal.

"Start as far in advance of the due date as you can," Anuskiewicz advised. "You can't be too early, because if you're too late your proposal won't be considered."

Without injuring reality, we could apply a paraphrase of the old horseshoe nail proverb: "For want of a proposal, the grant was lost. For want of the grant, the idea was lost. For want of the idea, future progress was lost." And if not permanently lost, it was seriously delayed. Individual businesses and the U.S. as a nation in an era of global competition cannot afford such delays when success hinges on development of new technologies and markets.

At your own company, neglecting to prepare and submit a proposal could delay or prevent an important development and significantly affect your future as well as the future growth of your business. This is not a new truth, but an old one resurrected for the 1990s and the approaching 21st century. It was already a fundamental scientific fact in the 19th century:

> "How different. . .would have been the history of our great inventors had they all possessed that knowledge of business affairs which would have enabled them to put their inventions in a business-like way before the world, or before the capitalists whose assistance they wished to invoke. The history of invention is full of illustrations of men who have starved with valuable patents standing in their names — patents which have proved the basis of large fortunes to those who were competent to develop the wealth that was in them."
>
> Scientific American
> December 18, 1880

These words from over a century ago speak loudly to the present. They dramatize the continuing obligation to develop and deliver successful R&D proposals, which in essence are simply good research and good business procedures put in written form with respect to particular projects.

Proposals to cover products, services, programs, and activities may be equally important and call for comparable efforts and commitments. Most can agree that a better choice than starving a good idea is doing whatever is necessary to advance it by preparing a forceful and functional proposal.

CHAPTER 3

ELEMENTS OF A PROPOSAL

"He who has hit upon a subject suited to his powers will never fail to find eloquent words and lucid arrangements."
Quintus Horatius Flaccus
8 B.C.

"The Wright Brothers flew right through the smoke screen of impossibility."
Charles F. Kettering

"It's easy to build a philosophy," said Charles Kettering, "It doesn't have to run." He couldn't say the same of a proposal.

A proposal does have to run, and fast, usually with a heavy luggage rack of complicated ideas. More often than not, it must run in competition with other equally ambitious proposals. Thus, the task is one of assembling the necessary components to make a proposal that runs and reaches the target ahead of others.

The job is more than simply presenting the facts you want to communicate in an orderly fashion. You must present facts persuasively. You must convince an expert reviewer that you know what you're talking about, that what you're talking about is an idea that deserves an affirmative response, that you're the best choice for the project under consideration.

Dashing off a proposal just to rush it into the mail so it can reach the desk of the recipient with minutes to spare more often than not in all likelihood is going to prove a waste of your time and your proposal reviewer's time. Whether or not you should bother to do the proposal at all is a serious question unless you can give it the necessary attention.

(?) Can you think it through in detail.

(?) Can you take pains to organize your arguments.

17

(?) Can you decide on specific objectives and specific ways to reach them.

(?) Can you arrange your facts in a logical, coherent order.

(?) Can you work — really work — to prepare a thorough proposal that persuasively communicates the conclusions you want reviewers to reach.

(?) Can you work on the language to edit out clumsy, foggy writing and oversee production of a tidy, attractive, inviting proposal.

If your answer to these questions is a reasonably unguarded "yes" or at least "I'll try," you're ready. You might remind yourself occasionally that a proposal is not a document to scribble off above the clouds between airports. Of course, if you've seen the movie, you can work on it — think about ways to refine the project, begin drafting some of the proposal elements.

A good proposal isn't written overnight. It evolves. And you can work on the different elements anywhere: Clarify their meaning and shape their phrasing.

Playwright Richard Sheridan roughly 180 years ago penned a classic warning for proposal writers when he advised: "Easy writing's curst hard reading." In other words, be ready to suffer some, Sheridan counseled. Most proposal reviewers faced with a stack to work through if confronted by "curst hard reading" are likely to hurry on to more palatable material.

Maybe the Golden Rule for those preparing grant proposals was stated by American Press Institute managing editors at Columbia University:

"Write unto others as you would be written to."

This rule lies behind the constant harping from idea inception to proposal submission on the importance of accuracy, clarity, realistic self-appraisal, methodical presentation, and clear speaking. Anyone who has tried seriously to arrange words in a proper order that succinctly communicates arguments and ideas to someone else already knows there is no substitute, except possibly genius, for diligent effort.

Historian Thomas Macaulay, a noted English stylist, had the audacity to assert what he called "the first law of writing": "That law to which all others are subordinate is this: that the words employed shall be such as convey to the reader the meaning of the writer."

Is that all? All you need do is put the right words in the right order to convey your meaning, and you have a proposal suitable for review.

Or do you? Even more is involved in a successful proposal than accurately conveying your meaning. The purpose of the proposal is to convince a government agency, foundation, company, or other organization that an investment of money should be made in you and your project. So you have to convey meaning and also make a sale. Submitting a proposal that does anything less is gambling against the odds and inviting rejection.

Making the sale could be even tougher if your ideas push forward into new territory, if they clash with established views, if they set off revolutionary fireworks. When you challenge the status quo, you must work that much harder to make your proposal a sterling example of the art.

Lee de Forest and Samuel Langley are representative of great scientists whose initial presentations to supporters and the public fell short. They needed stronger proposals to overcome natural resistance to the new and to eliminate conventional barricades.

Lee de Forest invented the tube that made radio possible, but he was tried on charges of fraud and ridiculed by a district attorney who accused de Forest of making the preposterous claim that "it would be possible to transmit the human voice across the Atlantic." The shocked D.A. told a court, "Based on these absurd and deliberately misleading statements, the misguided public has been persuaded to purchase stock in his company."

A December 10, 1903 editorial in *The New York Times* advised Professor Langley not to waste his time, money, and reputation in further airship experiments. "He is capable of services to humanity incomparably greater than can be expected to result from trying to fly," the editorial insisted.

The difficulty of putting a new idea across to a proposal reviewer can be overcome most effectively by systematic and diligent effort throughout the operation with particular emphasis on careful reflection as the first essential:

Think it through: Work on your idea until it is razor sharp. If you are tepid and only mildly enthusiastic about the idea, your proposal is likely to be the same and to convey a sleepy or hollow sound. Believe fully, even ardently, in your idea and this conviction will add authority and assurance to your proposal. With the idea focused and eager to go, you can begin assembling the different elements of the proposal.

Proposal Elements

Requests for proposals, RFPs, normally give detailed instructions concerning what elements to include in a proposal. You should follow all instructions exactly. A proposal submitted in response to an RFP is no place for "poetic license." Give them what they request or run the risk of having your proposal returned to you as incomplete or rejected out-of-hand as nonresponsive.

The elements required in a proposal may vary somewhat from one RFP to another. This fact makes the following rule a law as absolute as gravity: *Read the instructions*. Read them very carefully. Take notes. Then read them through again, and supplement your original notes about what is required. The RFP instructions are your recipe, and you can't afford to omit ingredients.

The 1991 edition of *The Foundation Directory*, containing proposal instructions and information for close to 7,600 foundations, describes a proposal as "a written application often with supporting documents submitted to a foundation or corporation in requesting a grant. Preferred procedures and formats vary. Consult published guidelines."

Applicants ignoring those varying procedures and formats run the likely risk of having their proposals sink into unfunded oblivion. After choosing a tentative government agency, foundation, or other recipient for your proposal, your advance planning and research should include consulting the potential target by letter and/or telephone and learning all you can about the organization's specific proposal requirements and instructions. In addition to details on procedures and formats, you'll have to know *how* many copies of the proposal to submit, *where* to send them, and *when* proposals are accepted.

Many proposals will be required to contain most if not all the following elements in the final proposal package:

A. Cover Letter
B. Title Page
C. Table of Contents
D. Proposal Summary or Abstract
E. Introduction
F. Statement of the Research Problem or Program
G. Objectives and Expected Benefits of the Project
H. Description of the Project
I. Timetable for the Project
J. Key Project Participants
K. Project Budget
L. Administrative Provisions and Organizational Chart
M. Alternate Funding
N. Post-Project Planning
O. Appendices and Support Materials
P. Bibliography and References

Some elements naturally will carry more weight in the review process than others. A technical proposal rises or falls on the strength of the problem statement, objectives/benefits, and project description. However, even when these elements are loaded with skillful grantsmanship and scientific sales appeal, the proposal still may fail if project participants are not demonstrably competent to conduct the project, if the budget is out of line, if follow-up financing is not available to keep the project going when grant funds are exhausted.

All the elements in the proposal package are not equal, but each is important and should receive careful treatment; and each element should be consistent and harmonious with the proposal as a whole. The elements should be neatly arranged and together should offer an attractive appearance. What the proposal says is the primary concern, but reviewers will find it difficult not to judge proposals in part on their appearance and graphic attractiveness. Books aren't supposed to be judged by their covers. Nevertheless, you wisely make your proposal look as good as you can, in a dignified rather than ostentatious fashion. Your goal in making the proposal neat and inviting is to achieve optimum readability with key elements and your main points coming readily and tastefully to the attention of a reader. This does not mean the proposal has to be expensively outfitted in costly binders, paper, and typography. Delicacy is required in this respect. You don't want to give the impression of overdressing the outside of the proposal because of concern about the contents. Good taste and simplicity are your best guides for the physical treatment of the proposal.

The main elements of a proposal are described below:

Cover Letter

The proposal instructions may not require a cover letter; but if they do not prohibit such a letter, use one with the copies of the proposal you submit.

The cover letter can introduce you, start establishing your credentials for the project, highlight special features of the proposal you want to underscore, and add any useful details not included in the proposal.

A cover letter may be less significant in the case of proposals submitted for grant competitions, but the letter is a vital and valuable part of the package with an unsolicited proposal. In the case of competitions when proposals are distributed to peer reviewers, a cover letter may not be seen by the reviewers. With the unsolicited proposal, however, the cover letter is the first thing read and commencement of your selling effort.

The type of proposal and the recipient's review procedures will dictate what you say in a cover letter and whether or not one can be used as a vehicle to help your proposal succeed.

Title Page

The title page or proposal cover sheet often is a form provided by the recipient of the proposal. Items commonly included on the title page are:

1. Title
2. Submitted By (Name and Location)
3. Submitted To (Name and Location)
4. Principal Investigator for the Project
5. Proposed Cost
6. Proposed Duration of the Project

A proposal cover sheet used in the Small Business Innovation Research Program is reproduced as Example 1. The small business certification data apply only to this competition and would not ordinarily appear on a title page. All the information requested on a title page must be given. One item in particular deserves a lot of thought—the title itself.

Just as a company pays close attention to the name given a new product, you should reflect carefully about the title for a proposal. If possible, your title should be brief and memorable while identifying the project and its benefits. The choice is critical, so don't leave title selection to the last minute and then settle for the first thing that comes to mind as a description for your project. That isn't the way to name a baby or a product. And that isn't the way to choose an attention-getting, mind-grabbing title for a winning proposal.

The title will become your project's public identity. Thus, the more impressive it is the better.

Table of Contents

If the proposal instructions specify a preferred format for the table of contents, obey without deviation and shape your proposal accordingly. To become a grantee, give granters what they want.

Proposal Summary or Abstract

Example 2 is a proposal summary form used in a state research fund competition. Like the title, the proposal summary is a prominent element demanding thoughtful preparation. The proposal summary is likely to be one of the last elements of the proposal that you write. It should evolve throughout the preparation process and state your case succinctly and well. The summary may be more frequently read than other sections of the proposal—a fact that justifies special care.

Example 1: Proposal Cover Sheet

This title page was required as a cover sheet by a federal department in the Small Business Innovation Research Program.

Proposal Title: _____

_____

Submitted By: Firm _____

Address _____

City _____ State _____ Zip Code _____

Submitted to: (Activity identified with the topic) _____

_____

Address _____

City _____ State _____ Zip Code _____

Small Business Certification:

The above firm certifies it is a small business firm and meets the definition stated in the Small Business Act 15 U.S.C. 631 and in the Definition Section of the Program Solicitation.

"The above firm certifies that it _____ does _____ does not qualify as a minority or disadvantaged small business as defined in the Definition Section of the Program Announcement."

The above firm certifies that it qualifies as a woman-owned small business firm
Yes _____ No _____.

Disclosure permission statement as follows:

All data on Appendix A is releasable information. All data on Appendix B, for an awarded contract, is also releasable.

"Will you permit the Government to disclose the information on Appendix B, if your proposal does not result in an award, to any party that may be interested in contacting you for further information or possible investment?
Yes _____ No _____."

Number of employees including all affiliates (average for preceding 12 months): _____

Proposed Cost (Phase I): _____

Proposed Duration: _____ months (not to exceed six months).

Project Manager/Principal Investigator	Corporate Official (Business)
Name _____	Name _____
Title _____	Title _____
Signature _____	Signature _____
Date _____	Date _____
Telephone _____	Telephone _____

For any purpose other than to evaluate the proposal, this data shall not be disclosed outside the government and shall not be duplicated, used, or disclosed in whole or in part, provided that if a funding agreement is awarded to this proposer as a result of or in connection with the submission of this data, the Government shall have the right to duplicate, use, or disclose the data to the extent provided in the funding agreement. This restriction does not limit the Government's right to use information contained in the data if it is obtained from another source without restriction. The data subject to this restriction is contained in pages _____ of this proposal.

Example 2: Proposal Summary Form

This form was used as part of a proposal package in a state research fund competition.

DO NOT WRITE IN THIS SPACE

PROPOSAL SUMMARY

Application Number:

1. TITLE OF PROPOSAL: MAKE IT BRIEF, INFORMATIVE, AND
                             WORTHY OF A WINNER!

2. TECHNICAL ABSTRACT:(Limit to 300 Words)

                        THE ABSTRACT SHOULD ACCURATELY
                        SUMMARIZE, PIQUE INTEREST, AND
                        STIMULATE A WISH TO LEARN MORE.

3. POTENTIAL FOR COMMERCIALIZATION:

                        GOVERNMENT GRANT PROGRAMS OFTEN
                        GIVE PREFERENCE TO PROJECTS WITH
                        STRONG POTENTIAL FOR EVENTUAL
                        COMMERCIALIZATION WHICH LEADS TO
                        TECHNOLOGICAL PROGRESS, ECONOMIC
                        GROWTH, AND NEW JOBS.

The proposal abstract may be published by the recipient of your proposal as a description of the project submitted for funding.

The length may be stipulated in the proposal instructions, with a limitation of 300 words as shown on Example 2 a popular length restriction.

If the instructions provided do not impose a limitation, you should restrict the summary to a single page.

The summary should include a clear statement of your project program, the research objectives, the anticipated results and benefits, and your qualifications to achieve the objectives. A successful summary will arouse the reader's interest and make him want to know more about the project.

Reviewers of proposals approach the task a number of ways, but one approach many reviewers favor is to read the abstracts in a stack of proposals first to gain an overview of what each proposal is about and awareness of the topics covered. Holding this background information in his mind, a reviewer then deals with each separate proposal in depth. This review method makes it obviously desirable for your proposal to be fronted by a forceful and persuasive proposal abstract that wins attention and favorably influences the reviewers.

Introduction

If the proposal format includes an introduction, use this proposal element to provide background information on the project and yourself. The introduction can help confirm your qualifications, experience, and resources to perform quality research of the type proposed. Such information may appear at length elsewhere in the proposal, but the introduction provides advance, capsulized support in these areas.

The introduction is a particularly useful part of an unsolicited proposal.

An effective introduction gives a few lines of history about your organization; and it briefly covers your goals and accomplishments, your successes in related research, your proven ability. The introduction prepares the reviewer for the sections of the proposal dealing with the specific project.

You might assume when preparing the introduction that the reviewer is unfamiliar with you and your organization, that he will know about you only what he reads in the proposal. This is a valid approach even if the proposal goes to a government agency, company, or foundation from whom you received earlier grants, since the proposal might be sent elsewhere for peer review.

Statement of the Research Problem or Program

Here we are coming to the heart of the proposal. This is an element that proposal reviewers consider with special care, because it specifically identifies the nature and thrust of the problem that the project will solve or it presents the rationale for the particular program intended to be carried out in the project.

The statement should establish a clear connection between the project and the organization submitting the proposal. Proposal instructions vary, and the statement of the problem element may sometimes receive other descriptive titles; but each proposal commonly is required to include such a statement and to relate it to the qualifications and experience of the firm seeking a grant.

A proposal in a competition for state grants to conduct an Appropriate Technology Dissemination Program was required to contain a "Statement of the Problem" which was defined as follows:

> State in succinct terms your understanding of the problem presented by this RFP.

The National Endowment for the Humanities included the following question among those considered by the reviewers evaluating proposals for Conservation Survey and Treatment Projects:

> Does the applicant describe the need for the project and place it within the institution's overall operations and priorities?

The statement of the problem is the place in the proposal to show a pressing need for the project and to link this need with the experience, activities, and history of your organization.

You need to be very clear about what the problem *is* so it follows logically and is equally clear what the problem *isn't*. If you overstate the problem, you may be giving those making the grant exaggerated expectations about what you intend to accomplish.

Identify the problem you will confront in your plan of action. Make certain it is a problem that realistically can be solved. Avoid grandiose prose and inflated description.

The winning proposal in the Appropriate Technology Dissemination Program competition mentioned above opened with the following statement:

> *Extensive information was generated during the Department of Energy Appropriate Technology Small Grants Program. Potentially valuable "how-to" technology publications and training materials will be distributed by the National Center for Appropriate Technology (NCAT). To achieve maximum utilization of NCAT items as well as appropriate technology aids from other sources requires effective and persuasive dissemination to organizations and individuals. This project will assess the appropri-*

ate technologies and pinpoint those with the greatest potential benefit . . . The heart
of the challenge is to accomplish the widest possible adoption of the technologies
economically and efficiently.

The winning proposal in its statement of the problem also summarized the experience of the organization in meeting comparable challenges and its prior involvement in such activities. This section of a successful proposal connects the problem, need, and organization so tightly and logically together that separating them should seem close to sacrilege. When possible the statement of the problem should be fortified with statistical data, support statements by experts or others in a position to corroborate the assertions made, and concrete examples tying the statement to real world needs and situations. The statement should communicate genuine concern and a sense of importance, so reviewers will not say "So What" but will say instead "This has to be."

Objectives and Expected Benefits of the Project

When the project is completed, what will the results be? You should list these results as the primary and secondary objectives of your project. Primary objectives are major goals, secondary objectives are specific components of a primary objective.

The objectives should be concrete, attainable results that can be measured and readily identified when you reach them. On a journey, when the objective is Cincinnati, you immediately know the moment the journey is finished: You're in Cincinnati. Proposal objectives need to be comparably exact, so you'll know when you're there.

Carefully stated objectives give everyone concerned, the granter and the grantee, a checkoff list for the complete project. The objectives can serve as points of accomplishment in the timetable to keep track of progress. The objectives may be given as precise tasks in the work plan.

Avoid vagueness when listing your project objectives. Avoid general allusions to broad, indefinite goals. Absolute specificity will focus and concentrate the project on definite activities with definable results. Also important is the protection you have against misunderstanding by the funding organization.

The old adage holds true that if you promise them the moon and the stars, they'll never be happy with just the moon. When preparing a proposal, you must be on guard to keep from saying more than you intend to say and making promises you can't afford to keep. Thus, in the planning phase before organizing the proposal, you need to formulate proposal objectives you can live with and successfully carry out in the time frame and budget allowed.

Nothing is won except trouble by promising too much. Reviewers are generally highly experienced people in their fields, and they tend to reject proposals with objectives that are unrealistic and out-of-line. Those submitting such proposals are fortunate if the reviewers do protect them from their own overzealousness, naiveté, or folly by rejecting their proposals. Both the funding organization and the organization receiving the funds have a common interest in setting up a project with clear-cut objectives. Then each can know what stage a project is in at all times, and there will be no basis for disagreement about when the project is finished.

The objectives can be stated specifically without making them austere and unmoving. The statement of each objective should include expression of the benefits that will come from its accomplishment. Objectives and benefits are inseparable twins. Objectives are the intended goals of the project; benefits are the advantages and reasons it is important to reach the goals.

Establishing objectives, ask yourself the following:

(?) What outcome should this project have. What are the specific answers to the specific problems we need to solve.

(?) Is this objective one I can reach in the time given and using the methods available.

(?) Can I meet this objective considering the budget.

(?) When the objective is reached, what are the benefits to the organization providing the funds. Will this satisfy them.

(?) What makes us the appropriate choice to undertake this project and reach this objective. Why is our proposal best.

(?) Have I said too much.

(?) Have I said too little.

Description of the Project

The statement of the problem, the objectives, and the description of the project are the three main elements of the proposal. The other elements are important, but they are supplementary to this tyrannical trio. These three are usually critical to the success of a proposal, and are given especially hard scrutiny by reviewers. If one of these elements falls short, the proposal is likely to fall short.

The description of the project should be considered the central element of the proposal. The description explains in detail exactly how you will achieve the primary and secondary objectives within the stipulated time frame for the project.

The description includes your work plan, methods and procedures, the rationale for the approaches taken, and quantitative projections of accomplishments to be achieved. The work plan may list each project objective as a task with an explanation of what will be done—and when—to perform each task and thus reach each objective.

Proposal instructions may sometimes link elements together in one section. Instructions for a "Proposal Description," for example, may include the statement of the problem, objectives, and work plan together. The following instruction appeared in an RFP for state research fund R&D grants:

Proposal Description—The proposal description must begin with a clear statement of the specific research problem or opportunity addressed and its importance. The remainder of the proposal description must contain the following information in the order listed:

1. Describe the technological field involved and the proposal's relationship to the current status of the technological area. 2. Provide a detailed description of what is being proposed including specific technical objectives of the research and development effort, including the questions it will attempt to answer to determine feasibility of the proposal. 3. Describe the work plan for achieving the specific objectives. The methods planned to achieve each objective must be discussed in detail. 4. Describe the anticipated results of the proposed research if the project is successful. 5. Describe the impact of the proposed product or process on job creation and retention. 6. Describe the applicant's plan for future commercialization. 7. Describe the targeted or anticipated market(s) including findings of relevant market research and provide references to the source of the market information.

This particular instruction calls for an omnibus proposal description, and each item should receive scrupulous attention. No item should be omitted. Proposal reviewers often use point scoring systems with every desired element weighted in accordance with the funding organization's wishes. You hurt your chances by neglecting to supply any morsel of information the RFP requests.

By contrast, the instructions for the Appropriate Technology Dissemination Program grant previously mentioned were as follows:

Management Summary—Describe in narrative form the management methods and procedures selected by your firm to complete the project as described in the RFP. Include evaluation and quality assurance measures.

Work Plan—Describe in narrative form your technical plan for accomplishing the work. Indicate the number of staff hours you have allocated each task. Include a

PERT-type display, GANTT chart, or similar time-related chart, showing each event, task, and decision point in your work plan.

In this RFP, the specific demands are not detailed and specific; but proposals nevertheless should be detailed and informative. If what you say can hurt you in a proposal, what you don't say can also hurt. The work plan must be sufficiently thorough to show that you have been careful in planning, that each task will be managed in methodologically sound fashion, and that time and budget factors are realistically served.

While indicating how the objectives will be achieved through your work plan, you should also identify who will reach them, emphasizing the strengths of the research team your organization has put together for the project. The equipment and resources of a special nature that are available to perform the various tasks should be described, especially any unique items that competitors might not be able to offer in their proposals.

If the methods to be used in the project are unusual or original, these details should be positively stated; and you should make clear why the methods are the appropriate ones for the job.

Stressing particular abilities, qualities, and features of your company and its staff that put you ahead of others in managing the project tasks should be woven subtly—but not so subtly that reviewers miss it—into the project description.

Rudyard Kipling's famous serving-men might be consulted to make certain you don't leave out matter of particular moment and importance:

"I keep six honest serving-men
(They taught me all I knew);
Their names are What and Why and When
And How and Where and Who."

Have these serving-men alert and on the job in your proposal, and you should do fine. The description does its job if it informs the reviewer precisely how and when and by whom each objective will be reached using the work plan you propose, and if it convinces the reviewer that you're the "who" to do it. The description does not succeed if it wins the enthusiasm of the reviewers and you the grant but with a work plan that exceeds your resources to perform it.

Thus, behind the project description is much reflection, planning, and analysis of the entire program and all its intermediate steps. The same care required to pinpoint the objectives is demanded for the description of the project before the proposal is prepared. Of course, determining the objectives and the work plan to achieve them is one process. Sometimes the objectives must be altered to fit the work plan that can be done with the equipment, facilities, methods, and staff available. The realities of the project are ultimately dictated by the feasibility of the work plan. The objectives must reflect what you can do, not what you would like to do in an ideal world with all things possible.

A frequent criticism of project descriptions in proposals is that they fail to be explicit, that they do not spell out precisely what methods will be used, what steps will be taken, and what schedule will be met in reaching objectives. The reasons for vagueness are understandable. The temptation is strong to avoid being tied down by specifics; but giving in to this temptation produces a proposal that is likely to strike reviewers as indefinite, speculative, wishy-washy, incomplete.

In the proposal description, you have to show that you have made up your mind, reached decisions, know what you're doing and planning to do. The objection about failing to be explicit is also made as a criticism about other elements of a proposal whenever the sin of vagueness is committed.

Every element of the proposal and particularly the description must be clear and explicit with exact details that reviewers will interpret as measurable events providing the basis for evaluation of progress and project results. This puts on you when preparing a proposal the obligation to:

(!) Make certain each element is fully developed and formulated.

(!) Make certain you weed out all evidence of vagueness and ambiguity.

(!) Read the description through several times to guarantee that it is unambiguous in stating the work plan and relating project tasks to project objectives.

(!) Explain why the project design and methods selected for the work plan are preferable to others and how they will lead efficiently to the objectives.

(!) Justify the plan of action as the most logical, effective, economical, and verifiable way to proceed.

(!) Identify your firm as the one best-equipped to accomplish specific tasks in the plan of action accurately and on time.

(!) Keep rewriting the description — and other elements — until the final draft is correct, direct, to the point, and meets your highest standards.

When you think the description really is finished, read it through impartially as if you were coming to it a stranger, and ask yourself if you would approve and fund such a project work plan. If not, keep at it until the description meets the sternest criteria, until it is convincingly explicit, well-documented, believable, and inevitable. Keep at it until it satisfies the meanest critic inside yourself.

Timetable for the Project

The timetable must show when each task will start and when it will end. The timetable is an extension of the objectives and the work plan, serving them as a timekeeper and lookout. The timetable entries commit you to carrying out the plan of action in a given time frame. Those funding the project require such exactness generally as a necessary means of monitoring the work and assuring its timely completion.

The timetable also serves you as a continuing check on progress. In a sense, the timetable provides a chronological map of the project made up of the points when each task will be started and completed. The timetable is a reference for everyone involved throughout the project.

Sometimes the timetable may start resembling a tyrant if accomplishing specific tasks within the designated period turns out to be a strain. That's why the timetable receives anxious and close attention during the planning of the project. With each decision about tasks and methods, you realistically appraise the overall time frame, or boundaries, of the project and the ramifications of the decision for each task. Failure to consider the demands various tasks will make on people, equipment, and facilities can make the timetable early in the project seem more like a work of fiction rather than fact.

The timetable schedule is generally depicted graphically in the proposal on a time-line display or chart. The proposal instructions quoted on pages 29–30 asked for a "PERT-type display, GANTT chart, or similar time-related chart, showing each event, task, and decision point in your work plan."

A PERT (Program Evaluation Review Technique) chart is a pictorial representation of planned activities in a project. The chart shows the interrelationships of activities and the priorities of their completion. The PERT chart consists of connected circles or other shapes with task names in the circles or numbers that refer to event identification listings in the chart legend. The circles are connected by lines, and the lines are scaled distances showing the lapse of time between one task and another. One inch may equal one month, for example. This chart is a complex representation of tasks, suitable for complicated, multifaceted projects which cannot be adequately displayed on simpler charts. If a PERT chart is not required by the funding agency in the proposal instructions, there may be no reason to prepare such a chart, which is difficult for the inexperienced to develop and for some reviewers to decipher, though it is an effective way to deal with an intricate and involved project embracing a maze of interconnected activities. A PERT chart often tends to resemble a scientist's diagram of a molecule showing a tangled hookup of atoms.

Example 3: PERT Chart Representation

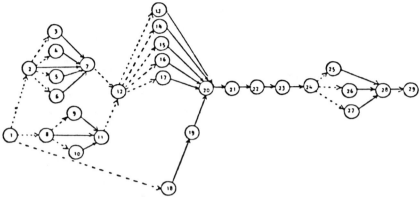

To learn about designing and utilizing PERT charts, see the following publication:

Desmond Cook, *Program Evaluation and Review Technique: Applications in Education* (Washington, D.C.: Superintendent of Documents, 1971).

Example 3 above is the diagram from a PERT chart used in a proposal. The numbers in the circles refer to events listed outside the chart. The solid lines are identified as lines signifying time-consuming tasks with dependency on previous events, while dotted lines indicate tasks that do not consume time but depend on previous events.

Fairly simple time-lines will normally suffice in a proposal to show when tasks and sub-tasks will be carried out. Example 4 illustrates two sample proposal schedules. Each is reproduced on the schedule form used in a state research grant program.

Several types of GANTT chart time-lines are available, but they are essentially alike in showing the start and completion of tasks and using different symbols to identify activities or requirements. Examples 5 and 6 offer samples.

Ideally the symbols used on a time chart are kept to a minimum. Each new symbol adds to the complexity of the display and makes it harder to interpret. When numerous symbols must be used because of need, confusion mounts and you run the danger of your chart starting to resemble a hodgepodge of competing marks. Time display mechanisms and the symbols used on a schedule should be the simplest available to meet the special requirements of the project.

Example 4: Sample Proposal Schedules

Two sample schedules are reproduced on the form used in a grant competition to illustrate the possibilities. On the form, S = Start of a Task, C = Completion of a Task.

PROPOSAL SCHEDULE

Application Number:
_____

List and number the major tasks and sub-tasks to be accomplished during the development of the product or process. For each task, show the start date, the completion date and any important intermediate milestones.

TASK	0	1	2	3	4	5	6	7	8	9	10	11	12	13	14	15	16	17	18
A.											MONTHS AFTER GRANT AWARD								
DEVELOP SPECIFICATIONS	S------------C																		
CONSULT USERS	S------C																		
MARKET RESEARCH/ASSESSMENT	S----C																		
DEVELOP SYSTEM		S----------------C																	
TRAIN PERSONNEL						S--C													
DEVELOP INSTRUCTIONAL MATERIALS						S--C													
B.																			
I. Purchase Materials	S------------C																		
A. Hardware	S------C																		
B. Chemicals			S----C															
II. Assemble Prototype					S----------------------C													
A. Build Model					S--------C													
B. Fabricate Final Parts							S--------C											
C. Final Assembly											S----C							
III. Test Prototype												S----------------C						

Whatever method is chosen to provide a running time schedule of the entire project, make certain that it shows as clearly as possible the following:

Example 5: 18-Month GANTT Chart

Each line runs opposite a specific task or sub-task (omitted from this example) of the original proposal from which this example was taken. The lines and symbols indicate the duration of tasks and the due dates for project reports.

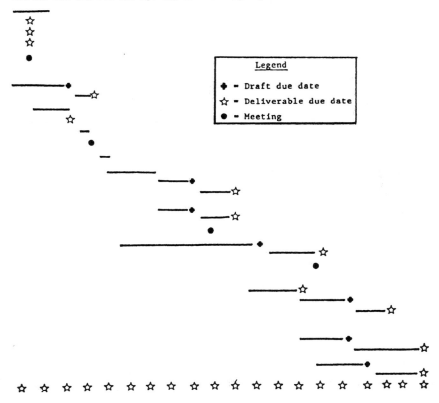

Project Months
Oct Nov Dec Jan Feb Mar Apr May Jun Jul Aug Sep Oct Nov Dec Jan Feb Mar

Legend

✦ = Draft due date
☆ = Deliverable due date
● = Meeting

1. Each task and sub-task listed
2. Starting and completion times for tasks and sub-tasks
3. Times when early drafts and final drafts of reports are due
4. Times for scheduled meetings, trips, reviews, and any other events that need to be formally entered on the chart

The timetable is simply a graphic representation of the project—a picture of the project from start to finish. This picture can help your

Example 6: Two-Year GANTT Chart Used to Show Duration and Scheduling of Tasks

Each line runs opposite a specific task or sub-task (omitted from this example) of the original proposal from which this example was taken. The lines and symbols put each task within a prescribed time period and identify when reports are due and when meetings will occur.

Chart Legend:
- ● – site visit
- ○ – meeting
- ■ – draft due
- □ – final draft due
- * – proposed program due
- Q – quarterly progress report
- S – site report

```
            First Year                    Second Year

F  M  A  M  J  J  A  S  O  N  D  J  F  M  A  M  J  J  A  S  O  N  D  J

   -●●●●

   ----*-■--□
   ------■--□
   ------■--□
   ------■--□

        --------

           —○-●●-●●-●●-●●-●●---
           ----S-SSSSSSSSSS-○----
        --Q--------Q--------Q-------■--□

                    -----■--□


                          ------
                       ---●--●-●--●-●--●-●----
                       --Q--------Q--------Q-----■--□--.

      ○                      ○                        ○
```

chances of succeeding with the proposal and also of succeeding with the project; but like many pictures, it needs to be touched up with care to assure that your project looks its best. Experienced reviewers of proposals often rely on the timetable when assessing the plausibility of a proposal, and they may be discouraged by a clumsy, hard to understand chart.

Investing extra effort to prepare an effective, neat, attractive timetable that explains itself quickly and conveys crucial facts about the project at a glance will sometimes pay off handsomely. A good timetable demonstrates your care and thoroughness, and also introduces vital clarity.

The Chinese proverb that one picture is worth more than ten thousand words isn't necessarily so (depends on the picture, depends on the words). But one good timetable in a proposal certainly makes much clearer several tightly packed pages of words. "A picture shows me at a glance what it takes dozens of pages of a book to expound," wrote the novelist Turgenev. Your proposal timetable can be organized and laid out to accomplish that function.

Key Project Participants

The identities, education, experience, and qualifications of the personnel involved in carrying out the project are always essential ingredients of a proposal. The funding organization not only must know what will be done in the project and what objectives will be reached, but also who will do the work. Proposal instructions commonly require those submitting proposals to include resumes of the team leader or principal investigator and any other staff people who will participate.

The length of resumes may be restricted by the instructions to a specific number of pages. Even if such restrictions are not given, applicants should avoid submitting massive resumes containing many pages of unsifted, unedited personal data. A pound of paper containing long lists of honors, publications, accomplishments is as likely to depress as it is to impress the reviewers, not to consider wear and tear on photocopying equipment. The resumes submitted with the proposal should be carefully edited to include only the experience and publications relevant to the project and potentially helpful in making an individual's qualifications quite clear to the proposal reviewers.

> (!) Don't leave anything out of the resume that pertains to the project and helps you establish your credentials as the right person for the job.

> (!) Don't include anything that is extraneous, irrelevant, and perhaps only ego-building.

A two to three page resume should work best, though sometimes longer resumes may be necessary for a particular individual or project.

Where the vitae or autobiographical materials for the participants physically land in the proposal depends on instructional requirements. Often these materials are put in an appendix and referred to when appropriate in the body of the proposal text.

The project description will identify who is responsible for particular tasks in the project. Such identification is especially worthwhile if a person of outstanding qualifications and perhaps distinction in the field will perform the task. Why a particular individual on a project staff is an ideal choice for a task should be explained and reference made to his resume in the appendix or elsewhere in the proposal. The proposal instructions may direct the placement of autobiographical materials in a separate section or subsection rather than in the appendices.

Many formats exist for the preparation of resumes or vitae to include in or accompany a proposal. Henry David Thoreau's dictum offers wise counsel here: "Our life is frittered away by detail . . . simplify, simplify." The simpler the format the better is a good rule to apply when compiling personal facts and achievements summaries for a proposal. The following items, selectively edited for relevance to the project, are normally found in the curricula vitae of R&D proposals:

1. Name together with address and telephone number.
2. Education — universities attended, degrees received, and fields of study.
3. Experience — work record with detailed descriptions of positions and responsibilities.
4. Publications — a list selected for project pertinence.
5. Professional Roles and Activities — include special assignments, editorships, presentations, workshops, and other involvements pertinent to the project.
6. Honors and Memberships.
7. Professional Skills — a documented description of the skills possessed by project participants will highlight their readiness and proven ability to perform the tasks and do the job properly.

If researchers in other organizations or universities will join the project as coworkers or consultants, their vitae should also appear in the proposal. Assuming you have a strong team assembled, don't keep the identities of participants a secret. Reviewers inevitably reach conclusions about the value, credibility, and prospects for success of a project on the basis of the people involved, as much if not more than on the objectives, methodology, and plan of action. People must implement the work plan and make it succeed. A plan is just a recipe of possibilities until the right people translate it into tasks accomplished. Reviewers need to know who those people are going to be.

Providing resumes or other autobiographical data on key participants should be an uncomplicated and headache-free part of the preparation process. Oddly though, this element sometimes turns out to be a serious, last minute roadblock. This happens, of course, because putting the vitae section together too often is put off until the end, possibly because it theoretically should be a simple matter.

(!) Don't put off assembling participant autobiographies until the other elements are complete.

(!) Set a specific deadline early in the preparation process when all resumes must be supplied by participants in edited versions complying with the agreed on format.

(!) If any participant ignores the deadline, get after that person with bloodhounds. Follow up persistently until all resumes are in.

A common practice in the hurry-up school of proposal preparation is to think about resumes in the last hours before the proposal must go in the mail or in the hands of someone charged with getting it to the right desk on time. Then fat, unedited, and generally out-of-date curricula vitae must be thrown into the proposal willy-nilly and an opportunity to strengthen the proposal substantially with carefully done personal summaries on participants is lost. In such situations, also, it is typically discovered that no resume materials whatever are available on one or more essential people who happen to be off on six-month field trips to Patagonia or Tasmania, and the persons with keys to their files have been missing for three days. They resurface one day too late.

(!) Don't take a chance.

(!) Start early.

(!) Get all resumes in hand during the opening stages of proposal preparation.

(!) Update, revise, retype vitae so they fit the proposal and look newly done rather than tattered resurrections.

If the resumes strike a reader as tired rehash jobs with no special relevance to the proposed project, the reader, alias reviewer, may unintentionally decide a tired rehash job could be made of the project. You can avoid such psychological booby traps by devoting a serious, first-class effort to the preparation of the sections that include participant descriptions. Remember that the credentials of project team members are heavily weighted in the funding decisions of most unsolicited proposals and have considerable import in grant competitions as well.

Ultimately everyone knows a proposal, no matter how well it is prepared, is only as good as the research team behind it, standing by to start and efficiently complete the project. So identify your team accurately, specifically, and yes, when warranted, enthusiastically. Probably no RFP has ever included the instruction: "Be humble." In your vitae, let the facts speak for themselves; but see to it that they speak clearly and loudly in compact, uncluttered formats that are conveniently brief, to the point, and easy to read. Some proposals fail because no pains were taken with this critical element. So be warned. Easy-does-it doesn't do it when new autobiographies are needed stat, to give your proposal a fresh, timely, and ready look.

Project Budget

What will the project cost? What are the direct costs, the indirect costs? Where is the money to come from? What is the contribution of the firm submitting the proposal? Are there matching funds available? What equipment is available, what will have to be purchased? Are the facilities you have suitable for the project, or will other facilities be required? What is the expected outlay for staff, support services, supplies, instrumentation, consultants, field trips, administrative needs?

These are some of the basic questions to be answered in the preparation of the budget, an indispensable element that may not make but well may break a proposal. The funding organization will be rightfully concerned about the economic feasibility of the project and will probably lose interest if it doesn't make sense financially as an investment. No source of R&D funds, whether a government agency, foundation, or private sector firm, wants to waste scarce resources on a project whose proposal budget shows it is too expensive for justification of the anticipated results.

The urge to make a good buy applies to providers of grants no less than purchasing agents, home managers, and teenagers with fixed allowances. Thus, proposal budgets receive probing scrutiny in the reviewing process. If the bottom line judgment in this appraisal has to be that the proposal offers too little for the money, the chances of acceptance will probably disappear or at least major revisions will be required.

You should also expect proposal reviewers to be highly experienced and thoroughly familiar with budget realities. If the budget to a knowledgeable reviewer appears inadequate for the proposed tasks in the project, the reviewer's judgment will have to be just as negative as when too much is requested. The funding agency wants its money's worth, and this is accomplished only with a realistically budgeted project that does not waste funds on useless frills but also does not cut corners or compromise the quality of the work to shave the budget a little closer to the fiscal bone.

A good way to make certain the budget is kept equitable and sensible is to work on it conscientiously from the start of the proposal preparation process. As objectives and the work plan are determined, parallel these determinations with running cost appraisals.

How much will this step cost in personnel? Equipment? Et ceteras?

Is this a cost-effective and task-effective step for the whole proposal?

Is there a cheaper way to go? Is the cheaper way good enough to reach the objectives?

Maintain a running budget as the planning proceeds. Put down the numbers as you go along. Don't guess! Figure the costs of each phase, activity, experiment, report writing, and conferences in Hawaii as exactly as possible. Put down the numbers. Obtain budget input from others on the project team who are closest to a particular task. Check their figures yourself and have others check them in the pursuit of accuracy. Keep putting down the numbers. This approach will become routine and fairly painless, and when you reach the last stages of proposal preparation the budget will already be written in the running record. And the budget will then come from the real world of work and money, not from last hours desperation to get the proposal done on time.

A glaring irony in proposal preparation crash efforts is how seldom methodical care is given to the job of putting the budget together. Again and again figures are dashed off, fitted into the budget form, and hustled out the door after a cursory glance and a goodbye kiss. And again and again proposals stumble and fall because too little reflection went into the budget. Or the budget succeeds along with the proposal, and pretty soon you wish it hadn't. You wish you'd stayed on that beach in Hawaii. You wish you were somewhere else, walking down a strange new street, working on some other project altogether, because you certainly can't break even with this one thanks to that recklessly naive, romantic, or insane proposal budget.

Exaggerated? Alas, not much.

Fiction writing, guesswork, inexactness may sometimes squeeze by in other elements of the proposal, but not in the budget. We're talking actual money here, not game money with another roll of the dice coming up. If budget errors do escape the reviewers' attention, you'll regret it when the errors come back to haunt you.

For instance, the U.S. Department of Energy conducted a nationwide Appropriate Technology Small Grants Program. Over $25 million was distributed in about 2,200 grants, funding the efforts of researchers to find ways of conserving energy and using it more efficiently. More effective solar greenhouses and sunspaces were one area of research. When the program was complete, the National Center for Appropriate Technology published a series of reports including, *Solar Greenhouses and Sunspaces — Lessons Learned*. This report contains the following comment which conveys loud echoes everyone responsible for compiling proposal budgets should hear and heed often:

"Underestimating the costs of contracted work when preparing a greenhouse or sunspace project is also a problem. Grantees in Connecticut planned to attach a greenhouse to a school. After the proposal was funded, the grantees discovered they could not find a contractor with solar greenhouse experience who could bid the job at the rate established in the grant. It is important to negotiate an estimated cost for the contracted service *before* acquiring the funding."

Is the message received? Is it loud and clear?

Example 7 reproduces the relatively simple and uncomplicated project budget form used in a state research fund competition. Budget items typically are supported by accompanying supplements or exhibits documenting the figures given. The exhibits required with Example 7 included:

1. *Personnel to Be Used on the Project*

   The annual salary rate, the wage rate, and the time needed on the project were required for each individual participant — personnel already employed by the applicant and personnel such as consultants to be hired for the project.

2. *Equipment to Be Used on the Project*

   A list was required covering the type of equipment, source, anticipated usage, cost, and whether the item would be rented or purchased.

The RFP for the state research fund, that used the Example 7 budget form, in a detailed explanation of direct and indirect costs noted: "The Department shall not permit an indirect cost or profit rate of greater than 15% from applicants, coapplicants, contractors, subcontractors, or consultants."

The items that will be allowed as direct costs are usually identified in the RFP with considerable detail. The entire RFP should be read, of course, with rigorous attention for fine points. The instruction you miss *can* hurt you. But the budget instructions call for rigorous attention plus. Direct cost items are those the funding organization pays for when funding the project. Direct cost items generally include:

Personnel Salaries/Wages

Equipment

Materials and Supplies

Consultant Services

Fringe Benefits

Travel

Facilities Rental/Lease

Special Direct Costs Determined by Negotiation

Example 7: Project Budget Form

This form was used in the proposal package for state research fund grants. The instructions said: "The application must contain a budget itemizing the major work elements and tasks to be performed including all direct and indirect costs to be incurred in carrying out the proposal."

PROJECT BUDGET

Application Number: _____

ITEM	State	Federal	Other	Applicant	Total
1. Personnel:					
a. Salaries					
b. Personnel Burden					
2. Consultant/Contract Services					
3. Transportation/Per Diem:					
4. Space Costs (Rent/Use)					
a. Space					
b. Office Equipment					
c. Office Furniture					
5. Machinery & Equipment Total (itemize on Exhibit D)					
6. Other Costs:					
a. Consumable Supplies					
b. Postage					
c. Printing & Publication					
d. Telephone					
e. Utilities					
f. Final Audit & Accounting					
g. Insurance & Bonding					
h. Memberships & Subscriptions					
i. Advertising					
j. Other					
TOTAL COSTS:					

ITEM ... SOURCE

Indirect costs, limited to 15 percent in connection with Example 7, are the costs to the organization making the proposal of providing a suitable site and services for conducting the project.

Example 7 provides columns for identifying the sources of funds to cover different items. This is necessary to define exactly what a project will cost and where the money comes from to pay the bills. Government and foundation grants are often made with in-kind contributions or matching funds required from the organization submitting a proposal. In-kind contributions represent various costs absorbed by the organization performing the project such as supplies, special instruments, staff time, and facilities.

Matching funds are a capital investment in the project by the organization seeking partial funding with its proposal. The availability of matching funds, even if not obligatory, can have a persuasive impact on the funding organization. Providing matching funds shows a strong commitment to and confidence in the project by the proposal maker, a cheerful willingness to put his own money where his project is. This is a reassuring fact to a funding agency which might with good reason entertain anxiety if the proposal in effect suggests: "Please invest/gamble that this project will succeed. I'm too nervous to bet on it myself."

The budget when submitted should be as finished and accurate as you can make it, although in some situations changes may be possible later. In the case of unsolicited proposals and some competitive proposals, the line items in the budget may be negotiable with the funding organization. Negotiating budget items is common practice; nevertheless, dismissing the necessity of careful budget preparation on these grounds is foolhardy. Successful implementation of a proposed project demands close and consistent harmony among the objectives, the tasks, and the budget. Even if the budget can be tailored later like a garment that is too tight or too loose to fit the project, achieving a proper fit before the proposal goes in is prudent, politic, and preferable. Show that you know what you're going to do and what it will cost by including a snugly fitting budget as a key element of your proposal.

Administrative Provisions and Organizational Chart

The proposal should identify specifically what administrative provisions have been made for the smooth functioning of the project. Explain special provisions made to guarantee efficiency and a friendly, inspiring environment in which good research must happen almost as a matter of course. Describe features of the organization that portray it as one where superior work gets done on time. You needn't mention the 24-hour per day availability of hot coffee in all labs, but it's a shrewd idea.

The proposal should promise regular progress reports and indicate when each report will be submitted. Provision for regular project review and evaluation should be made as well with details on how the evaluation will occur and when it will occur chronologically as the project moves forward.

The purpose of this proposal element is to give reviewers a clear idea of how you will manage the project, monitor the progress of tasks, achieve and keep up project momentum toward its objectives.

Alternate Funding

The project budget description provided above mentions the requirement for matching funds and in-kind contributions in some grant programs. Example 8 reproduces the introductory page of an application for financial assistance in a state research fund grant program. Note that in the section identifying funding sources, state, federal, applicant, and other sources are given space for designation of contributions. The applicant in the program was required to provide no less than 25 percent of the total funding.

The budget should give dollar figures for alternate funding, and the narrative sections should explain these provisions so the various funding commitments are thoroughly documented for reviewer understanding.

When a project's budget will be supplied from several sources, making appropriate prearrangements with authorized representatives of those sources is imperative before the proposal preparation stage is reached. Without firm agreements for alternate funding, there may be no need to prepare a proposal.

Has your firm agreed to put up the required matching funds?

Do you have solid agreements with other sources for their share of the funding to supplement the money expected from the target for the proposal?

If the answer to either of these questions is negative, you have more background negotiations and discussions to finish before moving ahead with your proposal. If you have reason to expect the agreements eventually, work on the proposal can take place to meet the deadline. But don't submit the proposal until you have the actual agreements reached—in writing.

All moneys tend to be equal in the sight of a fundseeker. You can look for alternate funding wherever it might be hiding in the public or private sectors. When your own organization has gone to the limit of the reasonable, other companies or institutions with a common interest in the project might be contacted for whatever support they have available—capital, staff, equipment, facilities. Their contributions whether in-kind or matching will help, and their participation as funding sources may add

Example 8: Application for Financial Assistance Form

This form introduced the budget section of a grant proposal. Note the required entries concerning funding sources and work schedule estimates.

```
                    EXHIBIT A                    ┌─────────────────────────┐
                STATE RESEARCH FUND              │ Application Number:      │
            APPLICATION FOR FINANCIAL ASSISTANCE │ SRFA 83-_____       │
                                                 └─────────────────────────┘
```

1. Applicant Name:_____

   Legal Address:_____

   Organizational Representative:_____
                                            Name

2. Telephone Number:__(___)_____

3. Proposal Title:_____

4. Funding Source
   Outline in detail from which sources the funds to be provided and estimated date
   that funds will be available.

		Amount	Date
a.	State Research Fund	$_____	xxxxxxxxxxxxxx
b.	Federal Funds	$_____	_____
	Name:_____ $_____		
	Name:_____ $_____		
c.	Applicant (at least 25% of Line 4e)	$_____	_____
	Source:_____		
d.	Other Funds	$_____	_____
	Source:_____		
e.	Total Funding	$_____	_____

5. Work Schedule Estimate:

   a. Beginning date of project work_____
   b. Completion date of project work_____

   _____   _____
   Signature of Organizational Representative of the Applicant   Date

   _____
                        Title

prestige and strength to the proposal. When funds come from multiple sources, you may have to assert yourself somewhat to retain adequate control of your project.

Post-Project Planning

Some proposals will involve one project only and have no requirement for later funding or plans for the project's continuation and future growth.

This indifference to the future is unlikely to be true of government agencies and foundations. In their acceptance of proposals, they tend to favor those that show promise of future life, further growth, and self-sustaining performance in the after-grant era.

Some grant programs, such as the Small Business Innovation Research (SBIR) Program which makes R&D grants to competing small businesses, emphasize the prospects for future commercialization as basic criteria for proposal acceptance.

Proposals submitted to win such grants should give evidence of careful thought and planning about the future and what will be done to assure project life after grant. Indicating sources of financing when the grant period expires is an effective way to demonstrate project durability. Giving details on objectives and tasks that will take effect when the objectives of the grant period are attained is a convincing technique as well to show that planning has taken place and that a strong commitment to the extended future of the project exists in specific terms.

Appendices and Support Materials

Staff resumes, organization history, product brochures, summaries of successful projects previously completed, letters of support and recommendation, amplification of proposal elements for technical specialists, and further documentation of materials in the main body of the proposal are examples of supplementary and support materials that may be included with the proposal as appendices.

The proposal instructions in the RFP for some programs will stipulate exactly what materials, if any, may be added to the proposal as appendices. Appendices that supply background information and documentation may routinely accompany unsolicited proposals. When preparing the proposal, discretion should be used, however, not to make the bulk of the proposal too overwhelming. Simply increasing the paper weight of an unsolicited proposal is not the best prescription for a robust and healthy proposal. Keep two rules in mind when deciding whether or not to add an appendix:

(?) Does the appendix add essential information the funding organization must know.

(?) If the appendix is omitted is the proposal appreciably weaker.

If you cannot give a secure and confident yes to each question, leave the appendix out since the probable reason it is considered has little to do with need and a lot to do with the misguided effort to make a proposal virtue out of thickness and heft.

Appendices are not allowed with some grant proposals. The Small Business Innovation Research (SBIR) Program restricts proposal length to 25 pages. This limitation is imposed so small businesses will be spared the expensive and time-consuming burden of preparing massive R&D proposals. The 25-page restriction also makes the review process for the SBIR Program easier and faster. The relatively brief proposal length in the program is a response to the frequent complaint by small businesses and individuals that they cannot compete in grant competitions with proposal writing staffs of corporations because of excessively complicated and demanding proposal instructions.

If the RFP instructions authorize the inclusion of appendices, you should judiciously and selectively take advantage of the opportunity. But think carefully about what is added and know exactly why it is added. An appendix thrown in for no good reason is a deadweight not a support.

Bibliography and References

If bibliographic and reference information is available and useful, include it to show the thoroughness of your research and your grasp of the technical background needed for the project. The references may be of a nature designed to show the special preparations and expertise your organization brings to the project.

This information can serve to identify noteworthy sources behind your statement of the problem, project objectives, and the planned research.

Many Elements Merge Into One

The foregoing descriptions give some of the features and facets of proposal elements. There are others, more than can be precisely covered, since each opportunity for a proposal is a little different from others and imposes special requirements of its own.

When getting ready to organize a proposal, the right attitudes are better than rules. All the elements summarized above will seldom be necessary in a single proposal. The elements you do use will have to be juggled for compliance with the particular needs of the particular situation. But the attitudes you bring to the challenge can always be the same from one proposal preparation onslaught to another.

What attitudes? The outlook needed when going into this particular battle is no mystery. Most of the old cliches about hard work, perseverance, sticking to it, and not cutting corners clearly apply. The requirements for the various elements of a proposal dictate what is needed and how to go about it. You can readily compile a personal list of pertinent attitudes, including some of these:

(!) A proposal is only as good as the preparations that go into it.

(!) Facts and specifics speak loudest and persuade first.

(!) Aiming at a specific target beats the scatter-gun approach — know where and to whom your proposal is going.

(!) You can't be too careful; you can't be too thorough.

(!) Last-day crash effort shows first-day neglect; start early.

(!) You can't start too soon.

(!) Faint heart seldom writes strong proposals or wins grants.

(!) A half-hearted proposal wastes everybody's time.

(!) Expect to win; commit whatever effort and resources you must to win.

(!) Proposal instructions are rules they mean you to follow.

If such comments sound rather like a coach's locker room exhortations to the team, so be it. Preparing a proposal will never be an easy or casual task. There's no logic in going through the experience unless succeeding with the proposal is a serious goal and a determined purpose. Care, perseverance, diligence may be popular words in Horatio Alger type slogans; but they have to be borrowed for the proposal preparation effort. Otherwise, don't bother. Go fishing instead. You'll have a better chance to catch something. Good proposals are the only ones that win grants, and good proposals are done by those who work at it.

Each element of the proposal may be separately prepared, but ultimately all the elements must merge together smoothly and harmoniously to create one element, the finished proposal. What Benjamin Franklin told John Hancock in another context at the signing of the Declaration of Independence applies to the proposal elements:

"We must indeed all hang together, or, most assuredly, we shall all hang separately."

If the proposal elements hang separately, the proposal is in trouble. They must flow, merge, and work together for a common purpose.

To achieve smoothness, coherence, and consistency in the finished proposal, one person should normally serve as the writer of the document. If a number of people make contributions to the contents, one individual should edit and if necessary rewrite what they provide to give the proposal the clarity and seamlessness that no group or committee could ever manage to achieve.

CHAPTER 4

GETTING READY TO WRITE THE PROPOSAL

"It is thrifty to prepare today for the wants of tomorrow."

Aesop's Fable of the Ant
and the Grasshopper

"The chief trait of the orderly mind is tenacity, concentration—that undeviating attention which in various sports is enjoined in the precept 'Keep your eye on the ball.' What we must keep our eye on in prose is the object, idea, or wording that we start with . . . For all who speak or write, the road to effective language is thinking straight."

Wilson Follett
Modern American Usage

Proposal writing is akin to genius if Edison was right that genius is one percent inspiration and ninety-nine percent perspiration. Proposal writing is ninety-nine percent preparation and there is generally quite a bit of perspiration involved. Consult the *Boy Scout Handbook* and you'll find a two word motto that offers the best advice possible for proposal writers: "Be prepared." Or look to Samuel Johnson for his sensible confession that "The greatest part of a writer's time is spent in reading; in order to write, a man will turn over half a library to make one book."

To make one proposal a proposal writer will necessarily turn over his company, several departments, books, all the experts he can find, and double the phone bill. Getting ready to write a proposal resembles preparations for a military operation. Staff must be alerted, operations orders must be issued, and a time designated for the invasion. Fixing the time is the easy part. That's the proposal due date.

The first and usually the hardest step is getting the inspiration.

That Indispensable One Percent

Edison may have said too little about the one percent inspiration ingredient in the composition of genius. Not much happens without that crucial fire to start things going. Uninspired perspiration, frankly, is just a lot of sweat.

Every proposal begins or at least should begin with an idea. Until the idea exists, there is no reason for a proposal and no apparent cause to work up that perspiration. But when the idea comes, the whole place seems alive and as Holmes used to tell Watson, "the game's afoot."

Dr. Raymond B. Fosdick, onetime president of The Rockefeller Foundation, wrote in *Chronicle of a Generation*:

> *"There is a common fallacy—and even some foundation executives may not be immune from it—that money can create ideas, and that a great deal of money can create better ideas . . . The difficulty is the lack of men with fertile spirit and imagination . . . with flaming ideas demanding expression."*

Good ideas generally seem to be born not made, and no one apparently knows exactly how it happens although there are battalions of books aimed at expediting the process from ancient times to modern. The Chinese philosopher Confucius recommended living "in the company of Men-at-their-best" which would provide a catalyst for excellence. Perhaps that is the same as saying "do your utmost." Confucius may also have known a little about preparing proposals. He noted that "if you have learned about System in the morning" by evening your task will be done and you can rest easy. The Chinese sage also said, "When strict with oneself one rarely fails."

The presses have kept busy in recent decades pouring out publications on the creative process and how to give birth to ideas—so you'll have something to write proposals about—from Ernest Dimnet's *The Art of Thinking* in 1928 to a cavalcade of contemporary books such as *Procrastination* by psychologists Jane Burka and Lenora Yuen who tell us how to stop agonizing, conquer our emotions, and reach goals. Exactly how we can be helped if our goal is an idea isn't quite clear. No figures are available about how many new ideas have been produced by taking advice on how to have new ideas.

Perhaps "think" is the magic word if there is one. An Italian proverb "Think much, speak little" may be useful in the preproposal stage of idea generation.

When he worked at Princeton University, Albert Einstein in the midst of tasks concerning the outer frontiers of knowledge often told his colleagues, "I must a little think." He would walk off alone and spend hours doing just that — thinking.

If there is a better formula than "I must a little think," it is probably simply an expansion of the idea for most of us to "I must a lot think."

Let's assume that thinking does the trick. You get an idea. It's a new idea, an appealing idea, and it seems to improve with age. You test its staying power with others qualified to judge and they admit you probably have something. The consensus is that your idea is promising and certainly should be checked out because "you may have something there." Checking it out calls for one thing:

You need to develop a project.

The Project

An idea isn't a project. The research project will take the idea and run with it. What does the idea need is the first question:

(?) Verification of feasibility

(?) Development of a product prototype

(?) Pure or applied scientific assessment

Turning an R&D idea into an organized, goals-oriented project calls for many and diverse professional skills. One thing is certain, if you can bring enough brains together to plan the project, you can also prepare the proposal that markets the project to a potential supporter.

W.I.B. Beveridge in *The Art of Scientific Investigation* made suggestions helpful in the process of translating an idea into a full-fledged investigation:

> *"A useful aid in getting a clear understanding of a problem is to write a report on all the information available. This is helpful when one is starting an investigation . . . Also at the beginning of an investigation it is useful to set out clearly the questions for which an answer is being sought. The systematic arrangement of the data often discloses flaws in the reasoning, or alternative lines of thought which had been missed. Assumptions and conclusions at first accepted as 'obvious' may even prove indefensible when set down clearly and examined critically."*

As the objectives and the needs of a project begin to take shape, critical examination by others as well as yourself will probably benefit the project. Just as few people can proofread what they write as well as some other person who can be more objective and less indisposed to find errors, so few scientists can troubleshoot their own ideas. Thus, consultation with colleagues and available outsiders is a vital part of the procedure when refining the problem to be studied and designing an effective project to reach whatever objectives are decided on.

Analysis of an idea to set up a project will include the following steps or variations:

1. Statement of the problem.

2. Consideration of ways to resolve the problem.

3. Consultation to identify the preferable approach to a solution and to specific objectives.

4. Designing a project to resolve the problem.

5. Consultation to refine and focus the project.

6. Determining necessary personnel, resources, and funds to carry out the project.

7. Identifying potential sources of project funding.

8. Evaluating the project's credibility, relevance, need.

Many projects inevitably fade away during the analysis stage. One way or another they fail to prove their worth. They prove technically flawed or too costly or unnecessary. Perhaps they seem lacking in general merit, timeliness, or credibility. They may be faulted on the grounds of originality or uniqueness. Rumors that others are aiming in the same direction have stifled initiative and discouraged projects before they leave the notebook and talking phase.

Bertrand Russell wrote, "Hardly any man of science, nowadays, sits down to write a great work, because he knows that, while he is writing it, others will discover new things that will make it obsolete before it appears."

If this threat hangs over an R&D project, the impulse in self-protection is to keep looking for a stronger idea, something all your very own.

The give and take of idea review and project construction does eventually run its course, and the result is a project that is needed, timely, beneficial to society and/or your company/university, cost-effective, and potentially productive. And also important, it is a project with highly desirable and realistically reachable goals.

So you have a project ready to go and a project team capable of performing the tasks involved. That is a situation calling for money, which means one thing:

You need to develop a proposal.

Furthermore, while you develop the proposal, you need to know precisely where the proposal will be sent.

Finding a Funding Source

When you have turned an idea into a feasible project and don't have sufficient funds to finance the project yourself, you look around for traditional funding sources: bank, venture capitalist, federal agency, foundation, private industry. Whatever source of funding is identified as your first choice (because it is your best choice), you will have to prepare a proposal directly aimed at the intended recipient.

A common mistake is to prepare a proposal without shaping it to the special needs and requirements of particular funding targets. A basic principle in marketing is directing the sales effort to a specific market, stressing the sales points that have the strongest appeal. The identical principle should be applied when preparing and submitting a proposal:

(!) Have a funding source in mind.

(!) Write the proposal to fit the needs and interests of that source.

Harry Woodward, associated with the Chicago Foundation, offered this counsel:

"Keep in mind that the foundation or federal program has goals of its own. Thought should be given to how your program will further these goals. It is often wise to be specific. Too often, organizations appear to be concerned only with what the grant will do for them and not on what it could accomplish for those individuals and organizations contributing the money."

Even for a nonprofit foundation or government agency with billions to invest, the proposal should allow the funding source to ask "What's In It For Me?" and easily find a satisfactory answer in the proposal. You should always expect organizations, the same as individuals, to make their decisions, including grant recipient choices, on the basis of the time-honored "WIIFM" (What's In It For Me) formula.

A generally organized and oriented proposal would not possess such specificity for a potential funding source. The classic marketing techniques should be employed as skillfully as possible. The proposal should show the funder how the project reflects and extends the organizational purposes and philosophical ends of the funding source.

The task of determining where to send your proposal is a research challenge in the case of unsolicited proposals. Many proposals, of course, will be direct responses to grant programs announced by government agencies and foundations. The decision about where to send a proposal should be based on exact, up-to-date, and thorough knowledge concerning the selected recipient.

Robert A. Mayer, when he was a staff member at The Ford Foundation, wrote in the *Library Journal*:

> *"The business of getting a grant has two sides to it: how to prepare yourself before asking for a grant and what the foundation staff member receiving your request will be looking for . . . How do you go about finding the right door? This is the next step, and a crucial one that many people seeking grants ignore. They will make proposals or requests to a foundation without finding out first if the foundation is interested. . . . In preparing yourself to ask for a grant, do two things: have a well-conceived, well-documented, hard proposal and know as much as possible about the foundation you are approaching."*

The same advice applies to the government agency, venture capital firm, private corporation, or other funding source you may approach: know as much as possible about their needs, interests, rules for proposals, and the types of projects they are known to fund.

Seeking a grant without bothering to acquaint yourself with the organization concerned is like threading a needle in the dark or having your proposal written by the six blind men of Indostan who fouled up so creatively in trying to describe an elephant by feeling different features.

Keeping Track of Government Grant Programs

Both federal and state governments sponsor R&D grant programs. These programs are publicized in various ways, and interested individuals and firms should make it their business to become expert in those ways.

Each department of the federal government (*e.g.*, Department of Agriculture, Commerce, Defense, Energy, Health and Human Services, etc.) can be contacted directly for information on current and upcoming grant programs available.

The *Commerce Business Daily* is a basic source on government business available by annual subscription from:

Superintendent of Documents
U.S. Government Printing Office
Washington, D.C. 20402

Subscribing to *Commerce Business Daily* or consulting it regularly at the library is recommended to those doing business or expecting to do business with federal agencies. The publication lists pending procurements of federal agencies and gives other pertinent information such as contract awards and surplus sales.

The Small Business Innovation Research (SBIR) Program which makes R&D grant funds available to small businesses which submit successful proposals is conducted by most large federal departments and agencies. In 1991 the following were participants in the SBIR Program:

Department of Agriculture
Department of Commerce
Department of Defense
Department of Education
Department of Energy
Department of Health and Human Services
Department of Transportation
Environmental Protection Agency (EPA)
National Aeronautics and Space Administration (NASA)
National Science Foundation
Nuclear Regulatory Commission

Each SBIR participant issues proposal solicitations well in advance of the due dates for receipt of proposals. The solicitations give details on the department's or agency's areas of interest and the technical categories for which proposals are sought. Copies of the solicitations are available on request to the organization concerned. Presolicitation announcements for the SBIR program and other government opportunities for small businesses and individuals are made by the U.S. Small Business Administration (SBA) to the SBA mailing list. To get on this mailing list contact:

Office of Innovation, Research and Technology
U.S. Small Business Administration
1441 L Street, N.W.
Washington, D.C. 20416

State grant programs and other state proposal opportunities vary from state to state. Contacting the governor's office and individual state government departments might acquaint you with grants that mesh with your interests, abilities, and special projects waiting for R&D funds.

Unsolicited Proposals

Worthwhile projects shouldn't have to wait around for a formal grant program to come along, and many projects needn't wait thanks to the unsolicited proposal option.

Unsolicited proposals are generally accepted by most of the organizations that conduct grant programs as well as others. Federal departments and agencies encourage the submission of unsolicited proposals by those with good ideas or services to sell that meet government needs and standards.

Most such departments and agencies on request will provide information concerning how to prepare and where to send unsolicited proposals for evaluation. The Department of Defense publication *Selling to the Military*, for example, contains a "Guide for Unsolicited Proposals."

Since any proposal has a much better chance of acceptance if it is prepared in accordance with the instructions given by the recipient, you should always take the trouble to learn those instructions before preparing a proposal for unsolicited submission.

A letter or a phone call can establish whether or not a private company, venture capital firm, or other nongovernment organization is interested in reviewing your unsolicited proposal. If such an organization is interested, you should find out what you can about the requirements and expectations of the organization concerning proposals.

Time is too valuable to waste doing a proposal the wrong way, especially since the right way usually can be determined with little effort. Also critical with unsolicited proposals is determining exactly to whom and where the proposal should be sent. If an unsolicited proposal is wrongly or carelessly addressed, the sender may never hear about it again.

The Foundation Option

As previously noted, the 1991 edition of *The Foundation Directory*, published by the Foundation Center, provides pertinent details, including proposal information, on close to 7,600 grantmaking foundations. The Introduction to the *Directory* indicates that these private and community foundations "awarded more than $7 billion or 93 percent of total foundation giving in the latest year of record."

Four types of foundations appear in the book: Independent Foundations, Company-sponsored Foundations, Community Foundations, and Operating Foundations. The foundations included are those with assets over $1 million or annual grant totals of $100,000 + .

The Foundation Directory is updated annually, and the 1991 edition includes 1,225 foundations that were not covered in the preceding edition. The publication, which is available in many public libraries, is valuable to "fund seekers, foundation and government officials, scholars, journalists, and others interested in foundation giving," or in short "everyone seeking any type of factual information on foundation philanthropy."

This *Directory* is one among many publications and services offered by the Foundation Center, a national service organization established and supported by foundations to furnish a single, centralized, authoritative source of information on foundation giving programs.

Complementing *The Foundation Directory* is *The Foundation Directory, Part 2* which offers equally thorough coverage on smaller but nonetheless important mid-sized foundations with grant programs between $25,000-$100,000. The book contains descriptive details on 4,200 foundations plus data on more than 25,000 recently awarded foundation grants. Such information can help you pinpoint specific foundation interests and thus determine where your proposals will have their best chances for grant-winning triumphs.

The Foundation Center is also the source of the *National Data Book of Foundations*, with over 30,000 U.S. foundations that make grants, the *National Directory of Corporate Giving*, covering more than 1,500 corporate foundations and direct corporate giving programs, *Corporate Foundation Profiles, Source Book Profiles,* a subscription service supplying timely details on the 1,000 largest foundations, *Foundation Grants to Individuals*, focused on specialized grant opportunities for qualified individuals.

The Foundation Center also publishes Directories, Grant Guides, Grant Indexes, and a wide range of guides, brochures, monographs, and bibliographies, all designed to assist grant seekers in preparing and submitting winning proposals to appropriate foundations.

The Foundation Center helps grant seekers identify and select the foundations "which may be most interested in their projects from the over 30,000 active U.S. foundations." Programs sponsored by and publications available from the Foundation Center assist each individual grant seeker in learning which foundation is the appropriate one for a particular proposal.

Foundation Center publications as well as information on the Center's programs and services, its network of cooperating libraries, and the location of the nearest library with Foundation Center materials are available from the Foundation Center.

The Foundation Center is located as follows:

The Foundation Center
79 Fifth Avenue
New York, N. Y. 10003
Toll-Free Telephone: (800) 424-9836

Foundations may be directly contacted for information on their programs and requirements. Important to remember is that if you have a strong idea and project ready to go, one or more foundations no doubt need you as much as you need them, because they have all that money ready to go. But don't fantasize that a benevolent foundation somewhere will issue a shovel and beg you to carry the money away. You need a strong project and an equally strong proposal to win a foundation grant.

Ultimately though you may discover that foundations are more receptive than either government or corporations in funding unproven, unconventional, radically new ideas. In a pamphlet published by The Foundation Center, F. Lee Jacquette and Barbara I. Jacquette described "What Makes a Good Proposal" and concluded:

> "Foundations are looking for more creative uses for their resources. If you have a project that really merits seed-money foundation support, push your plan—even if it deviates substantially from the general guidelines . . . We all have lots to learn."

Robert A. Mayer concerning grant opportunities offered by foundations suggested consulting *The Foundation Directory* regularly as well as other publications from the Center, noting that "These information sources will provide you with a better background against which you can frame your proposal, including the purpose and activities of specific foundations, the locale in which they make grants, and the general size of grants they make."

The broad scope of foundation grant opportunities as well as those provided through government and private sources shows there is no paucity of funders. It also shows that much work may be necessary to find out which source among the multitude is the best one to cultivate.

Usually the nature of the project will help reduce the multitude to a manageable few choices, though considerable study of the entire grant field may still be unavoidable.

Anticipate Now to Save Time Later

Whatever you can accomplish before going to work on the proposal itself will facilitate that struggle and its results. Among the advance steps you can take are:

1. Make certain all those you want as peer reviewers will be available and have them standing by. Each element and the entire proposal will benefit from careful evaluation by reviewers of your choice before it goes off to the funding source to run the gauntlet of the concluding and conclusive reviews.

2. List unanswered questions concerning the project—tasks, personnel, facilities, equipment, schedule, costs—and get all the answers possible.

3. Consult the funding source for verification of interpretations and resolution of questions about grant requirements and proposal instructions.

4. Set up a detailed timetable with delegated responsibilities for preparation, completion, and on-time delivery of the proposal.

5. Confirm budget estimates by contacting suppliers, consultants, external coapplicants, and others with expected involvement in the project.

6. Assemble documents that will add to or be part of the proposal — participant resumes, letters of endorsement, appendices, budget documentation, references.

7. Find out the personal work and travel schedules of everyone you're relying on for input and confirm that each will be present when needed.

8. If required, obtain the necessary approvals from your company executives and/ or department heads, both for the project and proposal and for the time of the staff involved.

9. If you can, identify others who have won grants from the funding source you're aiming at and interview them for any insight or guidance they'll provide.

10. Shift to another time any appointments that might interfere with or distract from total absorption in the proposal preparation effort once it starts.

Some of these steps may sound trivial or obvious. If so, you have all the more reason to take care of the matter in advance. The things everyone takes for granted — the obvious things — are too often exactly the things that give the most trouble because they are anticipated least. The unimportant becomes very important when it holds you up.

CHAPTER 5

WRITING TIPS AND ADVICE

"You write with ease to show your breeding,
But easy writing's curst hard reading."
Richard Brinsley Sheridan, 1751–1816

"Disordered speech is not so much injury to
the lips that give it forth, as to the dispropor-
tion and incoherence of things in themselves,
so negligently expressed. Neither can his
Mind be thought to be in Tune, whose words
do jarre; nor his reason in frame, whose sen-
tence is preposterous. . .Negligent speech
doth not onely discredit the person of the
Speaker, but it discrediteth the opinion of his
reason and judgement; it discrediteth the
force and uniformity of the matter and sub-
stance. If it be so then in words, which fly
and 'scape censure, and where one good
Phrase asks pardon for many incongruities
and faults, how then shall he be thought wise
whose penning is thin and shallow? How
shall you look for wit from him whose leasure
and head, assisted with the examination of
his eyes, yeeld you no life or sharpnesse in his
writing?"
Ben Jonson, 1572-1637

Playwrights Richard Sheridan and Ben Jonson penned their warnings about the importance of careful and precise writing long ago, but without knowing it they were aiming dead on at preparers of modern proposals.

Proposals that appear dashed off, that show no signs of mental sweat and writing care, are more than likely to be dashed aside contemptuously or indifferently in the review process. Clearly anyone who has occasion to put words on paper does his prose a permanent favor by keeping Ben Jonson's discourse perpetually in mind.

Applying Jonson's counsel to proposals, we could declare: "Neither can his proposal be thought to be in tune, whose elements do jar; nor his objective in frame, whose content is preposterous. . .a negligent proposal not only discredits the applicant, but it discredits the opinion of his project and judgment; it discredits the force and uniformity of the idea and purpose. . .How then shall he be thought wise — and worthy of grants — whose proposal is thin and shallow?"

Just as people regularly are judged by the company they keep and by how they look, the assumption should be seized and taken to heart that you will also be judged by the proposal you submit and by how it looks, sounds, and feels to those who must weigh, judge, and act on the specific contents of that proposal.

A proposal's recipients and reviewers, even if they know the person submitting it, still must reach their decisions based on the specific proposal that is put in their hands. So you can't afford to prepare a proposal with anything less than optimum care. You must package it effectively in the desired format and make certain it shows evidence of the clearest thinking you can manage as well as maximum "sharpnesse in the writing."

So Get to Work!

In most challenging situations including proposal writing, probably the best advice anyone can give another is simply: "Get to work!" Work is what it takes, and the only way the work of writing is ever accomplished is by getting at it. This work includes the conscientious pains you must take to put it right. "The time is out of joint," sighed Hamlet, the Prince of Denmark, and added, "O cursed spite, that ever I was born to set it right." If a proposal is out of joint, and it's your proposal, you had better do the work of setting it right before submission, that is if you're serious and expect it to fly.

We no doubt should acknowledge as well the sometimes painful necessity of buckling down to the hard and realistic work of thinking. That's the well-known secret of another essential ingredient in a good proposal — excellent contents.

Getting to work. Thinking.

If these sound easy when applied to the task of writing, try them some time. Both are demanding, tough, and often difficult. The temptation to put off the work of writing and the comparably intense work of thinking about *what* to write ranks among the least resisted temptations in the whole human Pandora's box of nagging temptations.

Mean deceit and cruel mockery are practiced when the claim is made that confronting a blank page, conquering the fear of filling it with words, and packing it with neatly ordered thoughts are elementary exercises for anybody, anytime, no sweat. Nonsense. Is it elementary to fight a shark or swim the English Channel? No way, nor is it elementary to do a proposal. The point is that you can with thoughtful effort learn what goes into a proposal and then with practice and work make yourself skillful at writing proposals that succeed.

Writing is seldom actually easy. Might as well admit as much up front. The waiting page really can be terrifying, and writing blocks are real adversaries the same as angry crocodiles, and that's no croc.

"To the vast majority of mankind nothing is more agreeable than to escape the need for mental exertion," wrote English statesman James Bryce, "To most people nothing is more troublesome than the effort of thinking." Add writing to Bryce's equation and like nervous gladiators we enter the arena of proposals, which require both persuasive writing and plenty of logical thought. In that busy and potentially lucrative arena, those who succeed must fight off the plague of inertia, which Van Wyck Brooks described as the place where, "The line of least resistance is to float. . .like a large inexpensive cake of soap."

Yet knowing there's a tough side to the task mustn't scare you away. Just the opposite. You have something worth proposing, don't you? You bet you do. The innovative project, idea, cause, purpose, goal egging you forward are on target and necessary, aren't they? Naturally. So why not turn to, get a move on, lights-camera-action, andiamo, start the work. Why not enter and fight it out if it takes till spring. Why not go ahead, meet that blank page head on, with head on straight, and do the proposal. Why not?

Blocking the Block

Maybe you have played or watched football and seen a brave, burly guard block a kick or a charging tackle. That's what we're after in the case of a famous writer's enemy known as Writer's Block. Holding ground and boldly blocking the block.

Writer's block is similar to job burnout or combat fatigue in that those who don't have the condition at the moment find it easy to disparage and shruggingly dismiss the condition in others. But those who actually experience writer's block — and most writers do at one time or another — know this is a tyrant capable of completely stymieing work and is a formidable adversary to be taken very seriously indeed.

When you have a writing block, you know. There's no mystery. Maybe you have this important proposal that is almost due, but you can't make yourself park and start. Writer's block is that vague dread, apathy, disquiet, emotional laziness inside causing you to postpone beginning the job as long as possible and justifying the procrastination with any excuse available. Writer's block, of course, is all in the mind and has an amazing talent for infecting the brain and fingers with a gripping paralysis.

Writer's block is when you fuss endlessly at sharpening pencils (you've been using a word processor for years now, boobie!), adjusting the light, fiddling with the thermostat, making another pot of coffee, calling your mother, calling all cars, doing whatever needs doing from chopping weeds, painting the fence, or reading *Anything* by Anybody simply to put off starting the word work ahead.

A writing environment in which you can work effectively, whether in an office or at home, is important. But don't allow the pursuit of a congenial environment to become a form of writer's block in itself. So much effort can be invested in finding the ideal place to write, you never quite get around to the actual writing that all the environmental and atmospheric fuss is about. Maybe tomorrow? As Vincent T. Foss noted, "One of the greatest labor-saving inventions of today is tomorrow." The fact is there is no ideal place to write. And if there were, writers would find some way or reason to hate it.

Writers learn that the only real antidote to writer's block is to force a start even if the results, hopefully on biodegradable paper, for many weary hours must land in the nearest circular aka rotating file. "You don't have to sharpen your pencils and sort out paper clips before you begin — unless it be your regular warming up," insisted Jacques Barzun. "Give yourself no quarter when the temptation strikes, but grab a pen and put down some words — your name even — and a title: something for you to see, to revise, to carve, to do over in the opposite way. When you have fought and won two or three bloody battles with the insane urge to clean the whole house before making a start, the sight of your favorite implements will speak irresistibly of victory, of accomplishment, of writing done."

Nearly all veterans of the great writing wars insist that victory over Goliath Writer-Block is best accomplished by brave, spunky, indefatigable little David Author through tackling the writing chores at regular hours, every day, whether in the mood or not. Amateurs, geniuses, and perhaps great lyric poets can wait for inspiration. The rest of us need to get to work and not kid ourselves that "inspiration" will make it easier. Hard work that inches forward, persistent efforts that make progress — those and only those are what make the task easier. Those and only those are what eventually will get the proposal written and the task done.

Each writer finds a personal way to start. Some make entries in journals. Some start the day with quick letters. Novelist and Nobel Prize Winner John Steinbeck prepared for his daily stints writing *The Grapes of Wrath* and other novels using both a journal and letters to break the block and start the flow.

Most who write, whether novels or proposals, learn the importance of isolation from others and the curtailment of distractions if circumstances allow. At the office, a closed door and the edict "No Calls!" can work wonders at blocking the block and letting you dive with a strong stroke into the heart of a proposal writing effort.

At home, unmistakable orders to the others outlawing interruptions must be delivered and, let us hope, obeyed. Writer Judith Krantz indicated that she posted this sign on the door to her writing sanctuary: "DO NOT COME IN. DO NOT KNOCK. DO NOT SAY HELLO. DO NOT SAY 'I'M LEAVING.' DO NOT SAY ANYTHING UNLESS THE HOUSE IS ON FIRE. ALSO, TELEPHONE'S OFF!" Now there's a "Don't Disturb Me & This Means *You*!" notice that succeeds at Krantz's house. Find out what accomplishes the goal of undisturbed privacy at your office or home and turn it on full.

Particularly if writing isn't a person's regular business, blocking the block and writing a proposal or other professional document may spell special trouble and call for even more discipline than a nine to five daily writer needs.

The occasional writer may lack confidence in his ability to communicate on paper. He may try too hard to imitate others rather than honestly express himself. Keep in mind: On a proposal that describes your idea and explains its advantages to a potentially interested recipient, you're probably the best writer in the world — after you jettison doubts, worries, inhibitions, and the I-can't-get-going blues.

Writing at the Laboratory or Office

Scientists, engineers, and business people who aren't regular writers (they probably should do more writing to get ahead) can do worse than learn from those who make a living, or try, from day-after-day writing. Daily writers routinely counsel "Practice, Practice, Practice" and "Persistence, Persistence, Persistence." They apply the stern rule that you learn to write by writing and become a writer by writing and are a writer when you write.

In the laboratory, university office, or business, for best writing results, write every day and at the same time if possible. Separate yourself from the daily hurly-burly to the extent allowed. If you have visiting hours, lunch hours, meeting hours, consulting hours, and coffee breaks, why not writing hours. "Permit no interruptions" was the number one canon and guide Max Shulman proclaimed as his key to writing.

Do your writing methodically. Treat it as serious work no less important than one-on-one sessions with colleagues, clients, and customers. "Plan the Work and Then Work the Plan" is a maxim some writing professionals live by. The plan-the-work dogma refers to legwork, research, what reporters Joseph and Stewart Alsop called "the rule of the feet." Working Washington D. C., they found that a successful reporter must use his feet as well as his seat. The first must move to get the story, the second must stay put to write it. "There's simply no substitute for diligence, for careful library work, for legwork and interviewing in the field of research. The 'rule of the feet' has no alternative in the craft of writing, whether it's fiction or nonfiction, professional or part-time," observed the Alsops.

Dorothea Brande in a 1934 classic, *Becoming a Writer*, said, "Writing calls on unused muscles and involves solitude and immobility." Producing a good proposal will give unused muscles a fine workout, and make the next proposal easier to do, just as the subsequent days of any hard exertion — gardening, volleyball, travel? — tend to be easier than the first days.

Brande was one of the "simply start working" advocates. "If a good first sentence does not come, leave a space for it and write it in later," she advised. "Write as rapidly as possible, with as little attention to your own processes as you can give." Brande's commandment was, "WRITE! Teach yourself as soon as possible to work the moment you sit down to a machine, or settle yourself with pad and pencil." She condemned daydreaming, reveries, desperate diversions, lackadaisical fussing. Grab your tools and jump in, she ordered: "Use the first sentence as a springboard from which to dive into your work, and the last as a raft to swim toward." Makes it sound like a blithe stroll in the country, doesn't she.

Doubts about the ability to write should be put outside with the rest of the family and cat. Editor, writer, teacher William Zinsser noted that one drawback to American education is the fear of writing it tends to inflict on many people. Zinsser stated, "Most people have to do some kind of writing just to get through the day — a memo, a report, a letter — and would almost rather die than do it." He insisted that professionals in all fields can and should develop writing skills and that the patient practice of actual writing brings inevitable progress in the craft of writing.

In a book aimed at easing the fear of writing, *Writing to Learn*, Zinsser pointed out that writing is a "basic skill for getting through life" and that it isn't an exclusive possession of "writers." Writing is the orderly and logical arrangement of thought that can be done by anyone who can think. "Writing is thinking on paper. Anyone who thinks clearly should be able to write clearly, about any subject," declared Zinsser. What does it take? Takes work.

Luckily the work of writing is also a superb exercise sometimes for the flow and development of thoughts. This is a fact scientists, engineers, and business people should memorize and apply to their daily tasks, problems, and proposals. Chemistry Professor Naola VanOrden insists that her students learn writing skills along with chemistry. "I believe that writing is an effective means of improving thinking skills because a person must mentally process ideas in order to write an explanation," she wrote. "Writing also improves self-esteem because mentally processed ideas then belong to the writer."

"Any time you reason your way through a complex scientific idea and put it in writing so that it's clear to somebody else"—which describes many proposal writing situations—you can feel good and should, William Zinsser asserted.

So *verbum sat sapienti est*, a word to the wise is enough, whoever you may be.

The Style is the Person

Writing style? You have plenty of writing style if you'll use it. Mark Twain gave a subtle tip on style when he said, "I never write metropolis for seven cents when I can get the same price for city." Another Twain insight that should help the wary and the worried was the observation that writing is simply arranging words in rows on paper and that all the words are in the dictionary waiting to be arranged. Back to Hannibal with you, Sam Clemens!

Style, frankly, is not something the proposal writer should lose sleep over or develop aches about. If you have something to propose and manage the task honestly, the style usually will appear. If you can achieve clear expression, that's a splendid style for proposals and most other forms of writing.

That is not, however, an invitation to dash off willy-nilly whatever comes to mind between the coffee break and lunch. We're after your *clearest*, your most *honest* writing here, which means careful, diligent work until you've done your *best* writing in your style. "Every sentence is the result of long probation and should read as if its author, had he held a plow instead of a pen, could have drawn a furrow deep and straight to the end," maintained Henry David Thoreau.

Rewriting, self-editing, more thinking, more rewriting, more research, further attention to the "rule of the feet," then still more rewriting. That's the secret of getting the necessary style out of yourself and on the proposal pages.

One of the literary Smiths, England's Sydney Smith, clergyman and wit, suggested, "In composing, as a general rule, run your pen through every other word you have written; you have no idea what vigor it will give your style." In proposals there's no need to be that vigorous.

In James Boswell's *Life of Samuel Johnson*, a book whose bicentennial came in 1991, there is better counsel for proposal preparation in what a tutor at Johnson's college told pupils: "Read over your compositions, and wherever you meet with a passage which you think is particularly fine, strike it out."

In a proposal, the style you want is the one that reaches and convinces your audience. Your job is not to entertain but to persuade. Max Shulman referred to the kibitzer always staring over his shoulder while he wrote. The kibitzer was the reader Shulman wanted to reach. "When I have to make a decision between my taste and his, I always yield to his," said Shulman. The reviewer may serve the proposal writer as a useful over-the-shoulder kibitzer in this manner. Knowing and heeding the audience are important for all who write and essential for the proposal writer.

The "reading over" and "striking out" approach quite often makes valuable contributions to a sharp, clean, clear proposal style. Concerning "style," English Professor Bergen Evans offered a practical insight: "A man's style of expression is latent in the first word he ever writes—or speaks. I don't believe a writer develops a style. I think he stops imitating other people and writes his own way; and that's his style." So be yourself, truly yourself, and you have all the style you need to write a farewell to the troops, a report to the Board, a letter to your love, or a proposal.

Beyond the False Start

False starts are commonplace, perhaps inevitable. They serve a useful purpose by getting the writing process underway. Also the false starts give you something to discard therapeutically when you finally realize you've really started, and the words you're putting down hold thoughts you want to keep and use in the finished proposal.

How do you know when you're past the false-start stage and gaining momentum on the real thing? "What is a false start?" asked Rudolf Flesch in *The Art of Readable Writing*. "It's a beginning that doesn't do what a beginning ought to do. Psychologists tell us that an effective piece of writing should start with something that points to its main theme. In other words, you must put your reader in the right frame of mind; you must start by getting him interested in what's going to come."

Recognizing the existence of the "false start syndrome" can provide a tool that helps the writer overcome blocks and make a true beginning. And about those pesky and vexatious writing blocks, remember that even world famous writers were not immune. One celebrated literary anecdote involves the writing blocks of Victor Hugo.

When the French master hit his inevitable periods of blockage, he disrobed and ordered his servant to remove *all* clothes from his room and not bring them back until a designated time. Thus, he would be a voluntary prisoner, naked and alone, with nothing but writing tools. Nude composition paid off, of course. Hugo gave the world *Les Miserables*.

This technique might help in extreme situations for unwritten, almost due proposals, even at the office. Hang your clothes outside the office door, and have them removed with understanding that you'll trade a finished proposal for their return.

Don't allow the false starts, or a bad first paragraph or page, to discourage you or drive you away. Accept them as familiar occurrences every working writer experiences. Accept false starts as a routine part of good writing. "Only the hand that erases can write the true thing" is the opening statement in Dag Hammarskjold's famous book, *Markings*. That's cogent counsel for proposal writers.

You should welcome the false starts as the vital overture to your real start and to the effective prose that follows on the page as you write.

Back to Work

We opened with emphasis on work. We better end with it. Urging writers to attack the waiting task, to start their work, is the ultimate tip of tips, the fond and friendly advice that counts most. The way to write is to write. The way to write a proposal is to write a proposal. Now.

If what others have done or do doesn't quite work for you, no matter. You're a person too. You're at liberty to find and apply your own surest methods.

The distinguished Civil War historian, Douglas Southall Freeman, had a daytime job. He composed his Pulitzer prize-winning biographies of Robert E. Lee working from 2 a.m. until breakfast. They say he took afternoon naps.

If that writing tactic sounds like a good idea, set your alarm. Or ignore Freeman's way and feel free to develop your own personal solution to the challenge of the right daily time, place, and private environment to carry on and accomplish your proposal writing job.

Right? Write!

CHAPTER 6

COMPUTERS MAKE THE TASK EASIER—MAYBE

"Man has created his own monster. He never realized when he invented the computer that there would not be enough statistics to feed it. Even now, there are some computers starving to death because there is no information to put into them. At the same time, the birth rate of computers is increasing by thirty percent a year. . .The scientific community invented the computer. Now it must find ways of feeding it. I do not want to be an alarmist, but I can see the day coming when millions of computers will be fighting for the same small piece of data, like savages."

Art Buchwald, 1969

"Computers are like anything else new or strange. People are reluctant to use them at first. They're afraid they won't understand how the technology works. But, once they see nothing bad happens to them and that their jobs are made easier, then people are more accepting."

Dr. Marietta Baba, 1991

Content matters most. Or should. In fact, the contents of your proposal ideally are all that matter. Ideally. But with that bit of obviousness admitted, we can add some reality to the equation. Since reviewers and others have to make decisions about the proposal package put in their hands, how it looks and the impression it makes inevitably count.

A book shouldn't be judged by its cover philosophically speaking; but realistically speaking, human judgment nearly always considers covers. Otherwise giant fashion industries would fold. They're thriving. Judging by appearances is an old human habit. Don't expect it to be put aside for your proposal.

That means the tools used for preparing proposals also matter. You want whatever tools make the process as facile, fast, efficient, and skillful as possible, and the final product as attractive, dramatic, and impressive as possible. Only fools and masochists make hard tasks harder by passing up easier methods of doing the job.

From the Quill to the Computer

The BHS (basic horse sense) Rule stipulates using the most advanced tools available that you are comfortable with and can operate with ease and skill to write a proposal. If you, or whoever is putting your rough draft into finished shape, are a novice on computers but a whiz on a typewriter, use the typewriter for the proposal in progress. Then give some thought to expanding your skills by getting friendly with a personal computer and one of the software programs such as WordPerfect or Microsoft Word for the next proposal effort.

In a former era, when proposal writers switched from goose quills to typewriters, they probably did so reluctantly. Switching tools is never done eagerly and seldom easily. Certainly newfangled typing machines had their critics. Robert Benchley declared, "The biggest obstacle to professional writing today is the necessity for changing a typewriter ribbon. Any school that can teach me how to do this can triple my earning capacity overnight. Anybody can write, but it takes a man with snake-charmer's blood to change a ribbon."

Now proposal writers, indeed most science and business writers, often hesitantly, sometimes painfully, make the switch from typewriters to computers. Lyric poets may be able to hold out against the tidal wave of progress in the technologies available for processing words. But they too will probably in time face a screen and a keyboard when the creative mood strikes. How easy is it to find parts and keep a beloved old typewriter going? And where can you get decent goose quills nowadays?

Contemporary writers using PCs may wonder on occasion if a little snake-charmer's blood might help as they struggle with whatever software programs they try to master. With a computer, the writer loses the personal control he enjoyed using a good hammer on the stone wall of his cave to chisel words and send his proposal to posterity.

Computers, especially to the uninitiated, can be scary and intimidating. That's the voice of computer anxiety we hear in *The New Yorker* cartoon where the little boy says to his computer hacker father, "Please,

Daddy, I don't want to learn to use a computer. I want to learn to play the violin."

Sad understanding comes when you lose an unprinted page into electronic nevernever land because you didn't save in the manner prescribed or hit the wrong keys. Then the melancholy and baffled sense of defeat surpasses that of the poet Coleridge when a notorious visitor from Porlock interrupted the writing of his famous fragment "Kubla Khan." When the amazing device, like a playful coquette refuses to obey, a quiet yearning for Benchley's old ribbon may creep in.

Thrust that yearning aside. The computer as a proposal writing tool extraordinaire is here to stay. After Henry Ford's Model T conquered the globe, the champions of traditional technology who shouted "Get a horse!" were ignored. In the same way and for the same reason, the computer is our new sovereign among efficient tools for science and business writing. The king's subjects simply must learn, relax, and reap the benefits. Good citizenship in Computerland starts with *learning*.

Master the Device—Or Use a Pencil

Initial emphasis should be on mastering this miracle worker, if you want to write proposals with a computer. An effective worker must focus on the work, not squander time and attention worrying about his tools and how to use them. Computer anxiety attacks during bouts of composition are fierce adversaries. For best results, the work of writing demands total concentration. Any distraction is an enemy to banish whether it's a pebble in the shoe or an unwelcome seizure of tool worry and processor fret.

Every tool has to be mastered and used with skill. With computers, the technology is complex: thus learning is more of a challenge and necessity than in quill, pencil, pen, and typewriter times. Then you could just start writing. Even on a typewriter, those adventurous cousins "hunt and peck" worked well enough if pokily. Computer technology is not as easy to pick up as quill technology, no matter what the computer magazine advertisements tell us. If your job is to write a proposal, you can't afford to be clumsy and hesitant with the writing tools you use.

Lincoln, the story goes, wrote "The Gettysburg Address" aboard the train to Pennsylvania. He could have done so faster perhaps and made it ten times longer if he had taken a laptop computer and been familiar with its tricks. But if Lincoln had tried writing his Address cold on a laptop without prior training, we can safely conjecture that Lincoln's prose poem would have come out differently after a series of "why the blankety-blank does it do that?" asides.

Yes, you can write a proposal on a computer, do a great job. But you have to know your instrument and your computer program thoroughly. If not, and if the due date is close, choose writing tools you can use calmly and effectively without computer anxiety, processor fret, or sheer paralyzed awe at all the computer options you're not skilled enough to use—this time. Don't start learning a new technology when you're facing a proposal deadline. That way you'll shortchange the proposal's contents and go bonkers to boot. Then in the quiet time after the original and copies are on their way, start getting yourself checked out on a computer for the next time.

How Computers Help in the Five Phases of Writing

"Computers should be a plus, not a pain" is the thesis behind a 1990s study conducted by Dr. Marietta Baba and Dr. Donald Falkenburg at Wayne State University. The thrust of the study, described in *New Science*, 1991, was to discover how computer technology can be introduced with minimum pain into technology-based firms.

Choosing computers, there is a clear need for user-oriented emphasis. What jobs do you have in mind and what sort of equipment do you need to do them? Wonderful features you lack the knowledge to use are like food locked in a safe to hungry people. "There are many documented cases of computer implementation failure which have resulted from a mismatch between employee needs and system capability," stated Dr. Falkenburg. "Computers provide a capability to radically change engineering and business practices," he observed.

"Computers are here to stay," Steven L. Mandell wrote in 1979, and he didn't exaggerate when he added, "It is therefore essential that people gain a basic understanding of computers—their capabilities, limitations, and applications." Mandell observed that businesses, government agencies, and other organizations rely on computers to process data and supply information for decision-making.

Over a decade later, Philip C. Kolin writing about computers made clear that Mandell if anything was timid in his prophecies. Few in the 1970s could imagine the dimensions of the computer phenomenon in the 1990s. Care to hazard a prediction about the nature and status of the clever gizmos in the year 2010? "The computer is an essential tool in the workplace," wrote Kolin in 1990. "Almost every type of business relies heavily on computers to generate, store, retrieve, and transmit information."

In proposal writing, the radical changes are external and physical, not internal. Using a computer doesn't affect the elements that go into the proposal or what is said. The computer does make some steps much easier and faster. The flexibility and versatility that are computer hallmarks can contribute in each of the classic stages that most writing must pass through to succeed:

The Five Phases of Writing:
Prewriting
Writing
Rewriting
Proofreading
Publication (Submitting the Proposal)

Some writers of fiction may skip a stage or two from indolence or impatience. Proposal writers live to regret it if they skip a stage or even cut one short.

Computers and Prewriting

Art Buchwald was only half joking when he wrote in his newspaper column about the insatiable appetite of computers for data and speculated about the need for a computer birth control program.

In the proposal prewriting stage when you conduct your background studies, perform library searches, and gather as much relevant data as possible pertinent to your topic, the computer is more than a Guy or Gal Friday; it's an everyday friend, pal, colleague, and speedy know-it-all.

Any organized collection of data put together for a proposal makes a database, even old standby 3″ x 5″ cards in an order that lets you extract facts quickly. Speed at pulling out data is the key attribute of a good database, whether on cards or computer.

The term "database" now generally refers to collections of data for computers that can retrieve the morsel of information you seek, when asked properly, fast enough to make Superman jealous.

In prewriting research, you should find out what databases the public library has available and use them. Many libraries subscribe to a computerized information retrieval system and offer on-line search services. Taking advantage of your database options can eliminate a lot of digging and provide information you probably wouldn't have without the computer's help. If there is a college or university library available, check there too. Also, businesses and corporations in your area may have databases and be willing to share information if you pay for the computer search.

When using a database, it pays to be clear about what you're after. So manually refine your search in advance. Seek advice from the reference librarian about what you need and how databases can assist. Pick your librarian's brains on the subject at a slack period when the reference desk isn't swamped with tenth graders, crossword puzzle addicts, etc.

Keep in mind that most government agencies now maintain valuable databases, and public access is encouraged. You won't be wasting your time to invest effort and learn all you can in this area. Here too the place to start learning what you need to know is the library.

A staggering and rapidly expanding amount of information is stored in databases and on self-contained disks that you can use for your proposal. Finding the source and asking the right questions are manageable challenges for you and your librarian, who is probably eager to help and a lot smarter than you ever suspected, particularly about databases, CD-ROM, InfoTrac, DIALOG, ERIC, BRS, WILSONLine, and other data gold mines waiting for information prospectors with proposals to write.

Computers and Writing

Those who started out writing in a paper culture and print-dominated civilization aren't comfortable at first with the idea of putting all their words and graphics on something called a "hard disk" or a plastic diskette, alias the floppy disk. But most writers after a surprisingly brief indoctrination period fall in love with the convenience and obvious advantages this means of writing has over all others.

The inescapable truth is that computers, used as they were designed to be used, can save you a lot of time, especially in the editing and rewriting stages.

Professional writer Jordan R. Young confessed that he started out resisting computers even though he was told often that word processors were ideal for his line of work. "Once I acknowledged their existence, I was inclined to agree," admitted Young. "I don't adapt easily to change — but I happily concede that computers are far superior to typewriters, when it comes to writing tools. Are you tired of writing and rewriting manuscripts, typing each successive draft? Do you often think of a more appropriate word — or several — when the final draft is in hand? Do typos distress you? Welcome to the computer era. With a word processor, you type anything only once. You can delete sentences, insert or alter words, correct typos, and even move whole paragraphs from here to there with extraordinary ease," Young wrote in a piece entitled "Plugging Into the Future."

Proposal writing profits from all the benefits Young listed plus the important area of graphics. Many computer programs now facilitate the production of marvelous graphics and graphic effects to dress up your proposal by making charts, work plans, time-lines, graphs, and budget forms efficiently, attractively, and fast. Some important material in your proposal may be easier to understand in graphic form, and computer graphics make it easy to present the material effectively in this way.

In the use of graphics, the choices should be determined by the nature of the proposal and its intended recipient. Fancy graphics in one case might be so out of place they would kill your chances of acceptance; whereas in another case, the fancier the better. A proposal to an Art Department about decor and the company's image calls for different graphics than a recycling proposal to the sanitation department. Know your audience and reflect that knowledge in your proposal's words and graphics.

You do have to resist the temptation of going too far with this facility for visual enhancement. Overdo it and bad taste sneaks in. The key to effective use of computer graphics again is learning the program thoroughly, gaining ease and confidence with it, so you can use it creatively and stay in control without running amok. Learn to drive the technology instead of allowing the technology to drive you. One proviso is to avoid using graphics for the sake of graphics, just because the capacity exists in the software. Don't use graphics merely to show off your fancy computer program. Make certain the graphics serve your proposal, not vice versa. Use graphics tastefully to enhance and clarify, not to create clamor and clutter.

Tip: Beware of clutter in your proposal and a sense of everything crowded in. Clutter and crowding can cause reviewers to feel claustrophobic and threatened. Provide plenty of white space to help readers feel safe and more receptive. Computer graphics, including the availability of different typefaces and sizes, give you strengths in this area that only existed at great cost in precomputer times. Now all is graphically possible right at your desk top, even your laptop.

Which computer software program should you use? Use the one you know, understand, and like. WordPerfect is popular in the business world. Microsoft Word has its champions as well, particularly in academia. These and other software programs and the computer hardware that uses them offer special word processing and graphics features to evaluate in terms of personal needs and inclinations. Software aspects to consider include the spelling check feature, the availability of a computerized thesaurus, graphics facility, accounting supports (e.g., can the program assist you with the proposal budget?), ease of use and editing, etc. You'll discover what matters to you when you get involved and determine the various options.

Where do you learn about computers? These days, anywhere and everywhere. Schools and companies offer day and night courses. Bookstores are crowded with computer books, magazine stands with computer journals. The trend of the times is clear when you find almost as many computer magazines available as sports magazines, and when computer columns in newspapers such as *The New York Times* are as popular and prevalent as entertainment columns. Libraries are major computer resources, and many libraries have computers available to the public. Videocassette rental stores offer videocassettes with software instructions.

Computer stores are as abundant as good delicatessens, and their staffs are generally more eager to serve. When you have several hours available, find a computer store clerk, and ask his advice and tutelage. In fact, you can teach yourself to operate a computer by following directions methodically, staying patient, and keeping cool. Learning computers is easy with many ways and places available to help. The main challenge will be to pause in the learning at some point and start using what you've learned to write another proposal.

Computers and Rewriting

Proposal editing and rewriting is the phase where your computer really shines since you can make changes directly in the copy without retyping the entire text. Additions, deletions, revisions, shifting elements around, switching paragraphs or sections — all are simple and routine on the computer.

However, be forewarned. When you change the existing text and replace it with the new text, you have the option of wiping the old text out forever. So think carefully before that happens. Take precautions:

- Remember when you tell the computer to delete something, it's *gone*!
- Have you saved a printed copy in case you change your mind again and want to replace something previously removed?
- Have you saved the text as it was on the disk in case you decide version one is better than the new, revised version?

Another facility the computer provides is the ability to use elements again, with changes as required, rather than retyping the whole thing each time. Thus, if you are preparing a number of proposals containing common elements, the computer delivers those elements instantly for each successive proposal and allows you to make any changes needed.

Proposal drafts can also be copied on disks and distributed to other proposal writing team members. They can make changes directly on the disks and return them to an individual who will evaluate all the changes and incorporate them in the final draft. This can add a new dimension of efficiency to the collaboration process or usher in total chaos. The team leader's job is to keep control and make the process work.

Computers and Proofreading

Computers work wonders as writing and editing tools, but ultimate accuracy is the writer's responsibility; which means the slow, essential chore of proofreading persists, largely unaffected by the use of a computer. Computer programs allow you to run a spelling check on what you've written, which is undoubtedly a useful support system to have and to use. But don't count on that pleasant service to be one hundred percent correct or to relieve you of careful proofreading responsibilities.

The proposal writer who retires his dictionary and thesaurus from active duty because the computer has a spell-check feature invites errors to visit and is certain to get them. You have to proofread carefully all proper names, addresses, and technical words. You have to catch those places where the wrong word is used, but correctly spelled—thus fooling the spelling check.

In short, the typewriter era task of proofreading is with us still in the age of computers. The proposal writer shouldn't lazily shift that duty to useful but unreliable computer software. The evidence increases that too many writers and publishers are passing the buck to their spell checks and failing to proofread competently. Such lapses in quality and standards should be avoided in proposals. If your competitors are careless about proofreading, that's their problem and your advantage. Word to the wise and the careful: Proofread!

Computers and the Finished Proposal

Here computers reassert their superiority. A variety of typefaces, typesizes, and other printing features allows the computer user to produce a high-quality, graphically-distinguished, finished proposal.

Take steps to achieve excellent, "letter-quality" printing for every element of the proposal. If possible, avoid submitting the proposal printed in the "dot matrix" mode even if the instructions say that is acceptable. If dot matrix printing can't be avoided, be certain the proposal is printed with a good, preferably new ribbon. If letters are weak and readers must strain, your proposal starts out with a strike against it.

If you don't own a quality printer for the submitted copy of the proposal, many commercial printing services now can take your disk and provide the high quality printing desired.

Should You Use a Computer?

The answer is definitely, if you know how and if equipment is available. Word processing and other data handling features make computers superb colleagues throughout the proposal preparation enterprise. Just don't become so enamored of the technology you lose sight of the goal in a showy but irrelevant electronic forest.

Remember the computer is just a composition device and order taker. It does absolutely none of the writing for you. You do all the writing and give the computer its orders. The technology benefits you most if you learn fully how to give those orders to obtain data, to construct appropriate graphics, and to preserve your writing.

Basic Tips for Writing Proposals By Computer

- Save your copy regularly, at least every 15–20 minutes, unless your word processor does it for you automatically. You may need to set a timer until the habit is established.
- Print often, identify successive drafts, and keep them in order for easy reference.
- Copy the contents of your disk each day on a second disk as a backup. If you send the disk to a colleague or a recipient, keep a copy yourself.
- Don't commit the computer sin of using features merely to show off what the device can do. Use its strengths to strengthen your proposal. Do nothing and include nothing that fails to serve the one goal of preparing and submitting a better proposal.
- Do know about and use the computer's remarkable agility and ability to paginate, alphabetize, form columns and tables, allow typeface variety, share the proofreading chore with a spell check, and many more jolly features you'll enjoy discovering.
- Don't pack every page full to the margins. Be generous with white space for the reader's sake.
- Do strive for letter-quality printing in all proposal elements. But don't kid yourself that fancy printing will let you get away with weak content. What you say is what really counts.
- Acquiring your own computer, don't rush. Haste makes waste and wrong decisions. Shop around, look, test, try, ask questions. Don't believe everything you hear or read. Pay attention, listen, compare. Study issues of the computer

publications—*PC Computing, BYTE, Computer Buyers Guide, PC World, Computer Shopper*, etc.
- Remember you don't have to be an expert and know every feature of a computer to use some features, the ones you need, very well. Concentrate on your needs and get really good at satisfying them with your computer.
- Never take the view that you can't work or write when you're away from that computer.

When the power fails or you're lying on a beach at Martinique, you can still work on the proposal. You can still write. Writing is the process of putting the words you choose in the order you want using whatever tool is handy. The computer makes it easier, improves the appearance, and expedites revisions, nothing more.

James Michener insists that all his best-sellers were written on the same old manual typewriter. Novelist Marcia Davenport at 87 was still writing on a typewriter she had used 45 years, because it was part of her habit pattern for writing. "In the first place, the keyboard suits me. Also, it makes a noise, and the clatter that it makes is part of the process. It feels right, and it's an extension of me. And a machine where the energy comes from plugging something into the wall instead of from here—the hell with it." So it was with Davenport at 87 and her 45-year old manual.

Michener and Davenport insist on their venerable machines, and creative power to them. But writing doesn't come out of machines, whether ancient or modern. It comes out of people. If push came to shove, Michener and Davenport could write on a computer; and you could do a proposal on their aging manuals; since content matters most. And the contents of the proposal come out of you.

CHAPTER 7

TEAM WRITING NEEDS,
STRENGTHS, AND PITFALLS

"A specialized writing job for a corporation is best accomplished by a team."
Robert Gunning
The Technique of Clear Writing

"Vigorous writing is concise. A sentence should contain no unnecessary words, a paragraph no unnecessary sentences, for the same reason that a drawing should have no unnecessary lines and a machine no unnecessary parts. This requires not that the writer make all his sentences short, or that he avoid all detail and treat his subjects only in outline, but that every word tell."
William Strunk, Jr.
The Elements of Style

Team writing, which can with discipline and effort be done well, is not the same as committee writing, which should be avoided whenever possible. The differences between a team and a committee are important. Teamwork occurs when a group of individuals with a common goal work cooperatively together to achieve the goal.

Committeework and the characteristic committee product more often than not are talk. Milton Berle described a committee as a group that "keeps minutes and loses hours." "I don't believe a committee can write a book," historian Arnold Toynbee observed on radio in 1955, "It can govern a country, perhaps, but I don't believe it can write a book." And a committee probably is no more effective at writing a proposal. Committee bashing isn't the purpose here. The point is that you need a genuine team, not a committee, for proposal writing. A committee typically starts to become effective when it evolves into a team, with members focusing on team results rather than individual prominence.

Walter Bagehot observed in his *Literary Studies* of 1879, "The reason why so few good books are written is that so few people who can write know anything." A proposal writing team does something practical about that problem by assembling a diverse group of talents and abilities and thus creating a pool of knowledge in the proposal's area of concern.

What has to be guarded against is having a team whose members are closet prima donnas, operating with committee outlooks and habits, while pretending to be a team. A successful team is fueled by cooperation not unruly and ragged individualism. Achieving and maintaining teamwork demand careful management and real commitment. A basic rule, therefore, is to choose team members who can perform as team players and serve the team, not their separate interests.

The Project and Proposal Writing Team

Yale Jay Lubkin in a 1990 article about "Getting the Contract" reported on a $1.5 billion contract won by IBM to upgrade the FAA air traffic control system. The winning proposal ran to nearly 50 volumes and was produced by an IBM team that included experts from many fields. The challenge clearly was beyond any individual's capacity, expertise, or time. Only a team and a large one could effectively respond to the FAA opportunity and have any hope of success.

"Once you decide to bid and have the necessary customer and product intelligence to back you up, the hard work isn't over," stated Lubkin. "Write the proposal from the top down. Establish a strategy. Set goals. Create your key databases. Storyboard the executive summary. Allocate sections of the proposal and write them. Above all, review, review, and review."

A project may be planned in terms of objectives and tasks, but it actually starts existing as a real instead of a paper project when you and those selected to work with you on the project are signed up for the work ahead.

The project team in many cases will also serve as the proposal writing team. This is a logical extension of team responsibilities, since the project team members have the greatest stake in assuring that the effort receives funding.

One of the first assignments for those on the team will be to round up the materials necessary for the different elements and sections of the proposal. This preproposal writing activity is indispensable for the achievement of a smooth process with no bottlenecks or barriers when actual proposal preparation starts.

The fewer facts that have to be determined, the fewer curricula vitae that have to be located, the fewer decisions that remain unmade the better, for a fast, untroubled proposal writing experience. Not to pretend, of course, that even such foresight will manage to think of everything and that no obstacles will rise up like specters to perturb and complicate the process. Confucius can recommend "System" in the morning, but it is a lot easier said than consistently done.

Proposal Writing Teamwork

When the proposal has multiple sections and a team is involved, sections of the proposal should be assigned for writing to members of the team with the skills pertinent to particular sections. The proposal writing team may include those who will participate in the project. It should also include, if available, anyone with experience in organizing and writing a proposal.

Whatever the size of the team, the designated writer (DW) — the best writer with the most experience preparing proposals — should have responsibility for bringing elements together from different team members, rewriting as necessary to achieve a cohesive whole, and in short achieving unified order among elements for the finished proposal.

The team coordinator (TC) task is to make certain that too many cooks don't spoil the broth, that the designated writer receives the necessary input and support, and that schedules are met. Or else. Since keeping the team inspired is a TC function, the team coordinator imitates Knute Rockne when so inclined and conducts or imposes appropriately timed pep talks.

The DW under the TC on the proposal writing team either produces the finished proposal personally, combining the submitted elements, or shares this duty with other writers who also hopefully possess experience in preparing proposals.

The entire team need not be involved in the proposal packaging, including the sequence of elements, the writing, the proofreading, the meticulous checking to be certain that all requirements are fully and accurately met. But all team members should review the results and enthusiastically but responsibly and constructively nitpick to correct errors, fill gaps, and improve the package.

On a larger scale such as the IBM proposal, the project manager and his subordinates must strive for similar cohesion and control. A lot can be accomplished to this end through advance planning, clear and precise guidelines, a straightforward agenda that all participants understand and follow, and a pattern of communication and cooperation among the different participants from start to finish. Otherwise, chaos.

The Proposal Writing Team (PWT) in Action

The team members should be selected as soon as the decision is made to prepare and submit a particular proposal. The "you, you, and you" method of choosing a team won't work here. The team members must be able to contribute to the advance research, the organization of relevant materials, the preparation of responsive elements, and ultimately the writing and/or review of the finished proposal. Since enthusiastic participants are preferable, qualified volunteers beat reluctant designees.

A willingness to work overtime and weekends to meet the proposal deadline is essential. A Little League play-off game at a crucial juncture isn't a suitable excuse for absence. That deadline date won't change even for Little League, or anything else. Imprisonment, paralysis, total amnesia, kidnapping, floods, earthquakes, and death have been given as excuses for failing to meet team commitments. None of these excuses are acceptable.

The full team should meet as early as possible, define its role, and make assignments. A team member should have specified responsibility for each element of the proposal. A realistic schedule of due dates for the accomplishment of successive proposal writing tasks should be set up and followed.

Each team member should receive at the start:

- Complete details on the planned proposal and its purpose.
- A list of individual team assignments.
- A copy of the relevant RFP or Solicitation to which the proposal responds.
- A schedule from day one to completion.
- Office and home telephone numbers of all team members.
- The company's project and proposal strategy including strengths to exploit and weaknesses to overcome.
- Guidelines for the preparation of proposal elements including exact format instructions.
- Completion dates for each element and review responsibilities.
- Follow-up team meeting dates and advance agendas.

The team coordinator must zealously ride herd on the entire process, see to it that scheduled events take place on time, and keep individual team members committed intently to the team's task and objectives. The team coordinator must keep his slack- and trouble-detector sharply tuned and focused so that ample warnings will signal weaknesses before they become serious impediments to progress.

The designated writer and his associates must avoid any ideas about outside activities during those busy hours when all the team members have their element materials in and the final compilation must occur. The whole team is on call throughout, while elements are prepared and during the final review and packaging phase.

A proposal writing team is no place, as Thomas Paine said about the American Revolution, for "the summer soldier and the sunshine patriot." Brave workers who do not shrink from the task or desert in a time of crisis are essential on a proposal writing team. The day the proposal is finished and submitted can be remembered by these veterans of the proposal wars and commemorated in later years as a memorial day, especially if the proposal wins.

You Don't Have To Be a Team

A proposal is a specialized task. For such a task, the contributions of a team are useful in assembling materials; and different professional skills can strengthen various elements of the proposal technically. Such skills are helpful, but you shouldn't assume they are essential. Good proposals that win have been put together by one or two individuals with no team whatever providing assistance.

Generalizing about what is and isn't necessary to prepare a successful proposal is reckless, since there are so many exceptions. The generalization does hold true, however, that the final writing group for a proposal should be one or two best-qualified individuals.

When a team tries to write as a group, the tendency to become a wrangling committee often intrudes to the extent that the team coordinator if surviving runs off to the South Seas. The danger can perilously balloon to the extent that the whole effort may collapse from a lethal overdose of agonized and disputatious argument. Of course, if the proposal isn't due until fiscal year 2005, you can let the group go ahead and try. There should be time enough later to fix the results, though it may be tight.

Team proposal writing brought IBM a $1.5 billion contract. Teams can put together skillful, comprehensive, successful proposals if they function as teams and not as yak-yak committees or anarchic herds of disgruntled captives. The proposal writing team simply needs to focus on the job, keep member egos calm and noncompetitive, and get on with the job — vigorously.

The team needs to reach the writing part expeditiously. Novelist E. L. Doctorow offered useful insight when he asserted in the *New York Times*, October 20, 1985, "Planning to write is not writing. Outlining. . . researching. . .talking to people about what you're doing, none of that is writing. Writing is writing." A proposal writing team, among a multitude of ancillary concerns, has to remember that writing is writing and that the team was born to write a proposal.

CHAPTER 8

PREPARE, EVALUATE, AND SUBMIT THE PROPOSAL

"To do our work, we all have to read a mass of papers. Nearly all of them are far too long. This wastes time, while energy has to be spent in looking for the essential points. I ask my colleagues and their staffs to see to it that their Reports are shorter.

(i) The aim should be Reports which set out the main points in a series of short, crisp paragraphs.

(ii) If a Report relies on detailed analysis of some complicated factors, or on statistics, these should be set out in an Appendix.

(iii) Often the occasion is best met by submitting not a full-dress Report, but an Aide-memoire consisting of headings only, which can be expanded orally if needed.

(iv) Let us have an end of such phrases as these: 'It is also of importance to bear in mind the following considerations . . .' or 'Consideration should be given to the possibility of carrying into effect . . .' Most of these woolly phrases are mere padding, which can be left out altogether, or replaced by a single word."

Winston Churchill, 1940
Memorandum on Brevity

The Prime Minister's commandment on brevity to his department heads might be communicated with benefit to the members of the proposal writing team. Proposal instructions often stipulate the maximum number of pages the submitted proposal will be allowed to contain. The instructions may even dictate the length of particular sections, such as 10 pages for the project description.

If the instructions simply specify the total number of pages allowed, those pages should be carefully allotted among the elements to give the elements that count most in proposal evaluation the greatest emphasis. The instructions for a state research fund grant program stated:

> *"The applicant may not submit a proposal of greater than 35 pages, elite type and single spaced, or 70 pages, pica type and double-spaced, not including attachments. All applications must be produced on standard 8 1/2" × 11" paper. Approximately one-half of the proposal should be devoted to the proposal description."*

This emphasis on the proposal description is typical. Reviewers expect other salient facts to be presented in compact formats — what Churchill called "short, crisp paragraphs." The pages on the details of the project are greater in number, because more facts, explanations, and discussion are expected.

If the proposal instructions do not provide any limitations on length, you should establish reasonable limitations of your own based on the requirements of the proposal and stick to them. The temptation to commit sins against the laws of brevity is too great unless firm specifications on length are given.

The individuals charged with providing materials for particular sections should receive length guidelines. If initial drafts exceed the final length intended for the proposal, no harm is done since later editing can remove the "padding" and compress the material into the length desired. Still editing "down" is never easy, so contributors to the proposal should know what the finished length will be and begin the distillation process early.

"To eliminate the vice of wordiness is to ensure the virtue of emphasis, which depends more on conciseness than on any other factor," argued Wilson Follett in *Modern American Usage.* "Wherever we can make 25 words do the work of 50, we halve the area in which looseness and disorganization can flourish, and by reducing the span of attention required we increase the force of the thought. To make our words count for as much as possible is surely the simplest as well as the hardest secret of style." Follett suggested that persistently editing our prose and making our words count helps sharpen a vital faculty any proposal writer should welcome, "the blessing of an orderly mind."

Brevity is the soul of wit, argued William Shakespeare, a man of many words who might have had problems organizing and writing an R&D proposal. Lee and Barbara Jacquette describing features of a good foundation proposal advised:

> *"Keep the written proposal short and clear. State at the outset what is to be accomplished, who expects to accomplish it, how much it will cost, and how long it will*

take. Avoid broad, sweeping generalizations . . . Test the proposition on others before submitting it. Be prepared to rethink and to rewrite. Learn about the foundations to which the request will go: be sure they are in fact operating in the area covered by the proposal."

Since unsolicited proposals generally have whatever length you give them, keeping the final result short and clear and unpadded is an internal responsibility calling for discipline of self and team from the start. In effect, if proposal writing instructions — such as length, elements to include, format — are not imposed from the outside, you should impose your own instructions on the inside to avoid the danger of a flabby, inflated, confused, and lopsided result. With elements of equal importance involved, a wordy contributor might fatten the proposal with 20 pages while a terse contributor starves it with 2 if you let them. Don't let them. Set up rules at the beginning and enforce them with reason and flexibility as friendly companions in the process.

In the absence of instructions from the funding source, contact the office charged with reviewing grant proposals and obtain whatever guidelines are available. If the funder provides any specifics about proposal length and other variables, use these as preparation instructions.

Don't Leave Anything Out

Whether you follow the directions given or those of your own deriving based on the situation, you should take precautions to guarantee that every necessary element is prepared and in timely fashion. Here are practical steps to take as work gets underway:

1. Start a prominently labeled file folder for each element to appear in the proposal. The folder will hold all materials produced on that element until the job is done. Develop the habit of automatically seeing to it that every piece of paper, including stray notes from volunteer critics, immediately goes into the proper folder.

2. Set up an element-person chronological display such as an enlarged version of Example 9. Put the display on a blackboard that won't be erased or other board that is prominent and will accommodate entries. List all the elements that will appear in the proposal, the person responsible for draft materials, and the due date.

   The last column should leave space to check off when the work is done.

3. Keep a similar chronological list alongside the master folders with columns for preliminary drafts and final drafts to be checked off.

When the proposal is done, the completed version should be carefully reviewed by the manager of the proposal writing effort and by other participants as well as uninvolved outsiders. Reviews for content should

Example 9: Proposal Work Chart

Job	Person Responsible for Job	Due Date	Work Done
Title Page			
Table of Contents			
Proposal Summary			
Introduction			
Statement of Problem			
Objectives/Benefits			
Project Description			
Project Timetable			
Project Participants (curricula vitae)			
Project Budget			
Administrative Provisions/ Organizational Chart			
Alternate Funding			
Post-Project Plans			
Appendices/Supports			
Bibliography/References			

also include, even if it seems obvious and elementary, an element by element check to make certain nothing has been omitted.

Assigning responsibility for elements by name and in public with accompanying dates should help expedite the task. No one likes to hold up the works, especially if this means being conspicuous about it.

Team members need to know and believe with piety that due dates are total despots and absolute tyrants put in power by the forces beyond, and they cannot be changed.

Write the Best Proposal You Can

You're hearing nonsense when you listen to the popular balderdash that writing quality, style, clarity, and good prose don't matter in a proposal. It's the project itself that counts, insist the lazy, the ungrammatical, the careless. Heed them not. If your proposal isn't well written, clear, and precise, the reviewers probably will never understand how good your project really is. Your job is to make certain they understand, which calls for all the proposal writing skill you can muster or hire.

"Writing maketh an exact man," promised Francis Bacon, and exactness is much prized when writing a proposal.

The poet Alexander Pope offered a bit of specific advice on the subject of writing, and his advice is worth repeating: "The secret of writing well is to know thoroughly what one writes about, and not to be affected."

Robert Gunning repeated this idea when he said clear writing results "only from clear thinking and hard work." He concluded, "We are all in the same boat. We need constructive criticism." Hear, hear.

Criticism starts with self-criticism and with the assumption that shoddy writing enormously reduces and perhaps erases your proposal's chance for success. At a minimum, proposal writers should acknowledge the warning issued by Edmond Weiss in his book *The Writing System for Scientists and Engineers*: "Every sentence you write runs the risk of being read."

When preparing a proposal, you can't afford to write just to fill the page, even though it has to be a certain length for dignity's sake and you can't think of much more to say. In that case, more thinking and more work may be needed. No one promised that writing a good proposal is a summer lark. What it is actually is an opportunity for demanding, sometimes difficult work; and if the work succeeds, the payoff is funds for that project you have your heart set on. This chance for success inspires those who decide to do the necessary work, and those are the ones whose proposals do the job.

Virginia White in the book *Grants* wrote the following:

> "The only aspect of the entire grant-getting process totally under your control is writing your proposal. Fashions in grants come and go as in everything else. What organizations clamored to support yesterday goes begging today because of national and international crises, natural disasters, or political vicissitudes. There is nothing the grant seeker can do to change that. But no matter what the political or social climate, or what the current 'glamor' field may be, every applicant has it in his power to prepare and submit a first class proposal . . . The most beautifully prepared application will not turn a pedestrian idea into an inspired one, but the best idea in the world, inadequately described or unimaginatively presented, can be misinterpreted or even overlooked."

White quoted a foundation official who noted the problems foundations have detecting "real or potential merit" in proposals "presented so poorly that it is easy to miss their worth."

The message of such statements is loud and clear: Be clear. And do the hard tasks of thinking and writing that lie behind clarity and make it possible to say clearly and factually what you mean.

In the 1970s, an analysis made by Louis Masterman of more than 700 proposals rejected by the U.S. Public Health Service showed the following rejection pattern:

Reason for Failure	% of Total
Nature of the project	18%
Competency of applicant not shown	38%
Inadequate planning and carelessly prepared applications	39%
Miscellaneous	5%

Poor proposals were implicated directly in 39 percent of the failures. The 38 percent attributed to the nondemonstration of applicant competency was also largely due to poor proposals. Most of the applicants were essentially qualified for the projects they proposed according to Masterman's findings. Their proposals simply didn't make those qualifications clear.

These results have been and are repeated by most grant programs. Proposal reviewers for government and foundation grants normally find that at least half of the proposals they receive are rejected as poorly done or nonresponsive. A government grant program reviewer was quoted by Virginia White in *Grants*: "We really had to scrape the bottom of the barrel to come up with enough recommendations and even then we were unable to approve as many as there were funds to support. Fifty percent of the applications had to be thrown out—they were hopeless."

Those who prepared the "hopeless" proposals did a lot of work for nothing. The sad truth is that many of them probably didn't do enough work, that they stopped short of giving the proposal preparation job an all-out, unconditional, never-say-die effort. Since writing proposals for nothing sounds like a good example of utter foolishness, a genuine going-for-it effort should be made or forget it until another project appears in which you do believe enough to prepare the best possible proposal.

If you have doubts that the best proposal you can write will be good enough, still you can try, and without going into the effort psychologically defeated before you start. No one knows your project better than you, and that depth of understanding makes you the best person to describe it fully and with exact details.

Among professional writers, poets are the ones who make themselves artists of exactness with the exactly right words in the exactly right places. It pays to listen to poets on the subject of exactness and being specific, because those are primary needs in a good proposal. William Blake gave good rules for proposal writing when he said, "He who would do good to another must do it in Minute Particulars. General Good is the plea of the scoundrel, hypocrite, flatterer; for Art and Science cannot exist but in minutely organized Particulars."

You alone understand the Minute Particulars of your project. You alone can organize them minutely for persuasive presentation to reviewers. If the actual writing of those particulars is something for which you have too little experience, get help with that part of the challenge. If your firm lacks someone with technical writing experience, such a person could be brought in on a temporary basis to participate. The added expense would be justified if you produced a substantially better proposal.

But chances are you and your proposal writing team can do the job quite well if you agree from the start to work on it hard and to keep going until you're convinced it's a winning job worthy of the grant you're after. If a proposal you submit doesn't win, don't let it be because you delivered a proposal that fell significantly short of the best you can do.

Proposal Evaluation

The most significant evaluation of your proposal will be the one made by the reviewers for the funding organization. Perhaps equally important, however, is the final evaluation made by you, team members, and other peer reviewers before the proposal is sent on its fund-winning way. Until it passes your internal evaluation, it shouldn't leave. This means a proposal should be in final form a sufficient time before the due date to allow an adequate review process. "Review, review, review," emphasized Yale Jay Lubkin in his 1990 article "Getting the Contract."

Entirely too often proposals are finished late and sent off without internal evaluation or even thorough proofreading. Guard against this costly haste at the end by setting up and insisting on a realistic work schedule, including reviews, when the work commences.

Evaluation is an essential part of the project itself, with provision in the project description for regular assessment of progress and achievement of objectives.

Proposal evaluation is a variation on this project design provision. Proposal instructions ordinarily give the evaluation criteria that will be followed for proposal evaluation. You can apply these criteria to the proposal yourself while it is being written and to the finished product before it goes out the door.

Each program will have different criteria, though similar types of programs may use similar grounds for proposal evaluation. The evaluation criteria in the proposal instructions should be studied carefully. If an unsolicited proposal is prepared with no instructions available, you should learn what you can about the evaluation procedures and criteria used by the funding source concerned.

The evaluation procedure given in a state research fund grant program began as follows:

> *"Proposals received by the department before the close of business on the deadline date will be reviewed and evaluated according to the following procedure: Proposals will be evaluated on a competitive basis in a two-stage process. Applicants will be pre-screened by the department to ensure that proposals comply with statutory requirements, rules and application instructions and have technical content and merit and potential for commercialization . . . In addition to the general criteria, the feasibility panel and other reviewers will review and evaluate the proposal based on the following specific criteria."*

The general criteria cover the eligibility of the applicant and his demonstrated capability to carry out the project as well as the commercial promise of the project and its benefit to the state economy. Twenty-two specific evaluation criteria are listed related to applicant qualifications, proposal content and merit, potential for commercialization, and evidence of other financial support. The proposal instructions include an Evaluation Scoring Chart, for use by reviewers, showing how criteria are weighted. Example 10 reproduces this form in part.

The points indicated for different criteria offer valuable guidance concerning the contents of the proposal. If 40 points are given for commercialization potential, you will find occasion in the proposal to discuss that potential as thoroughly and optimistically as possible. In such a grant program, if your project has little current potential, you won't bother to write a proposal. You'll look elsewhere for friendly funds.

The importance of knowing the evaluation criteria and the basic purpose of a grant program is fundamental to guide you in deciding whether or not it will pay to submit a proposal. The evaluation criteria give concrete standards to use in writing and then in reviewing a proposal. If you can achieve constructive criticism through the internal review process and correct proposal shortcomings, the reviews by the funding source should be correspondingly more favorable. Since self-criticism is not easy when it concerns work you're close to, objective peer reviewers should be used as well.

A point scoring system comparable to the one in Example 10 is fre-

Example 10: Evaluation Scoring Chart

The following chart, designed to be filled out by proposal reviewers, was reproduced in a grant RFP.

PROPOSAL EVALUATION SCORING SHEET

Proposal Name: _____

Application Number: _____

Reviewed By: _____ Date: _____

Grant Request Amount: $_____

CATEGORY   *POINTS*

I. Potential for Commercialization (maximum 40 points)

SUB-TOTAL _____

II. Content and Merit of Proposal (maximum 30 points)

SUB-TOTAL _____

III. Other Technical and Financial Support
(maximum 20 points)

SUB-TOTAL _____

IV. Application Qualifications (maximum 10 points)

SUB-TOTAL _____

TOTAL SCORE _____

(Note that the sheet contains the proposal name but not the applicant's name. This again suggests the importance of choosing the proposal title carefully.)

quently used in competitive grant programs, and variations may be applied in judging unsolicited proposals.

Funding sources show considerable variation in their proposal reviewing methods. In the federal government, some departments such as Defense review proposals mainly within the department. Others utilize elaborate peer review systems designed to achieve objectivity and fairness in judging and to assure that the best projects are singled out and funded. As an example, the Department of Energy in the Small Business Innovation Research Program used a step-by-step review procedure that included:

1. First step technical review.

2. Full scale peer reviews.

3. Point scoring by the technical topics manager.

4. Final selection of winning proposals based on point ratings and program factors.

You should find out all you can about proposal review procedures currently in effect — how they're reviewed and by whom. This information, applied when writing the proposal, helps focus efforts. The National Science Foundation for instance employs a rigorous peer review system that strongly emphasizes scientific and technical merit in judging proposals. A National Science Foundation spokesman said, concerning proposals from small businesses or individuals lacking scientific prominence in their fields, that the proposals will benefit if they include the names of respected consultants as project participants. A representative of the Department of Energy said concerning proposals from small businesses or relatively unknown individuals that a research connection with a university would add extra legitimacy to a proposal.

These are facts to learn about the funding sources to whom you consider sending proposals and to take into account when organizing projects and preparing proposals. Evaluation criteria are the points you'll be tested on. To pass the test with your proposal, you should demonstrate how your project and the evaluation criteria harmoniously come together.

Submitting the Proposal

This is the easy part and yet how often has it been fouled up. In grant competitions, many proposals — perhaps excellent jobs reflecting considerable work — fail because those submitting them neglect to follow instructions. They submit too late, to the wrong place, or in the wrong format.

You should go through the proposal instructions section that covers submission and the section on preparation of the proposal physically. Make accurate notes.

(?) Where do you send it.

(?) How many copies.

(?) Exactly when is it due.

(?) What packaging is required.

(?) How printed—what kind and size of paper, printed 2-sides or 1-side, single-spaced or double-spaced, margin specifications.

When the proposal is finished and presumably ready to submit, go item by item through your checkoff list of these matters and make certain a good proposal doesn't fail merely because carelessness took charge at the end of the process. On this issue, Yale Jay Lubkin in 1990 asserted, "Proposals are like everything else. If you want the job done right, it has to be organized and scheduled. The frenetic pace of most proposals, starting slowly and building up to a vast crescendo on the day before the proposal is due, and then collapsing into exhaustion the day after, is not the way to win."

At a seminar on the SBIR Program, Ann Eskesen, President of Innovation Development Institute, stressed the importance of proposal appearance and readability. "Make your proposal good to look at as well as read," she said. "Content is everything, but presentation is extremely important to make certain that content is seen and understood . . . Mechanical mistakes such as front-page typos, incorrect pagination, and silly errors should be avoided. This is important because you are asking for funds to do detailed work; yet you haven't even competently proofread your own proposal. This creates a harmful image."

To avoid this harmful image, do yourself the great favor of going to work on the proposal well enough ahead of time to do a thorough job and still comply with the proposal due date. Allowing a comfortable margin of time to finish a proposal days in advance of the deadline is good advice but often ignored advice. An easy way to be kind to yourself, prevent anxieties, and submit a better proposal well before the last minute arrives is obvious and routine: start earlier. Many shrug off this obvious and routine counsel of good sense, try to dash off a proposal with the last hours breathing down their necks, submit something less than their best, and often live to regret it. If you think you need a minimum of three weeks to do the job right, give yourself six weeks.

Fight the temptation to be prematurely satisfied. "If you are never satisfied with what you write, that is a good sign," argued writer and teacher Brenda Ueland in *If You Want to Write*. In the perpetual struggle to give it your best shot, there's no law against "persistence, persistence, persistence" as a corollary to "review, review, review."

How cost-effective is it to rush through and fail when you could take a little longer, work a little harder, make the appearance of your proposal worthy of its contents, and produce a winner? Mull it over.

If multiple photocopies are submitted, be certain all copies are clean clear, and sharp — not weak and smudgy. Appearance won't improve the contents of your proposal, but neat appearance will enhance the impression of care, seriousness, and professionalism you give through your proposal to the reviewers. Such an impression helps your chances. Don't think it doesn't.

CHAPTER 9

WIN OR LOSE, TRY AGAIN

"When you've completed one proposal, get to work on the next one."
Program Manager
National Science Foundation

"Debriefings should always be requested. Establish credibility, show them you're interested, find out what you did wrong. It may turn out there simply wasn't enough money to fund your project or that a particular DOD program has changed direction because of world events."
Program Manager
Department of Defense

Proposal writing is an art that practice and experience make better. The realistic expectation is that you may need a period of proposal writing apprenticeship before you hit your stride. Research organizations with considerable experience and expertise prepare and submit a lot of proposals, but they don't expect to win them all. In fact, a win percentage of 25 percent is considered respectable.

Naturally, disappointment is inevitable when considerable work is done on a proposal and apparently to no avail. But even if the proposal doesn't win the grant it aimed at, the work need not have been in vain. Often the same proposal can be modified, improved, and sent off again to another funding source — or even the same source — with better results because the proposal is better, or the timing is better or both.

The teamwork invested to plan a project and the proposal for marketing it is an asset that can be applied to other projects. The materials generated for the proposal often have spin-off uses in other proposals, company promotional literature, and technical stimulation of related projects.

The worst mistake would be to let the project and proposal team break up because of disappointment if one proposal doesn't hit. Zealously guard against letting disappointment drag you into the morass of discouragement which can quickly degenerate into the disease of apathy and inaction. "Discouragement is the only illness," George Bernard Shaw argued. A better move than quitting is to plunge the same group into another concerted effort, either to refine the original project and proposal for another go or to take off in some new direction. Try again is the only logical option.

When a proposal is completed and submitted, months generally elapse before the results are announced. During those months, you can simply go about business as usual waiting for the results. Or you can get busy on the design of another project and preparation of another proposal. Then whatever happens with the earlier proposal, you have something new in progress. The smart choice seems obvious.

The Value of Debriefing

Most government agencies, foundations, and other organizations provide a debriefing opportunity to applicants whose proposals are not accepted in a particular program. Whether the debriefing is on a group basis or an individual interview, you should make certain to take full advantage of the chance to learn specifically why your proposal wasn't a winner. Such information will teach you postgraduate lessons in how to avoid the things you did wrong and to do even better the things you did right when you prepare more proposals.

Yale Jay Lubkin went further. "Even if you win," he wrote, "get a debriefing so that you can do better the next time."

Refusing the debriefing because of resentment at not being selected is nonprofessional and harmful to your own future chances. From the viewpoint of your continuing education in preparing winning proposals, when a proposal wins you need to know specifically why. When a proposal loses, your need is even greater to know specifically why.

(?) Did you do something in the proposal you shouldn't have done.

(?) Did you omit something you should have included.

(?) Which element of the proposal was a problem causer and why. Or which elements.

The debriefing can help you answer these questions and others relevant to the proposal writing process. If you made mistakes, without debriefing, you could repeat the same mistakes in subsequent proposals. At a 1989 Small Business Conference on R&D Opportunities, SBIR Coordinators from various federal agencies repeatedly emphasized the crucial importance of persistence and the value of debriefings for those submitting SBIR proposals. The debriefings help establish contacts and build relationships between businesses and federal agencies that foster greater understanding of agency needs and serve as catalysts for winning proposals in later competitions.

When a good proposal doesn't win, and some of them won't, patience is needed with the process and perseverance to give it another try. Just as accidents will occur in the best-regulated families, according to Mr. Micawber, rejection will occur in the best-prepared proposals, sometimes for no reason other than that the topic arrives too late—or too soon.

You wouldn't take up a sport such as tennis and expect to defeat a champion immediately. The competition for most grants is intense, attracting champions and beginners as contestants. The same as in a sport, the beginners would be naive to expect early victories over more experienced competitors. But you can rapidly gain experience and earn promotion from beginner status by writing and submitting proposals on every project you and your colleagues can dream up. Practice will gradually lead to better and better proposals, and better proposals will bring grants.

CHAPTER 10

GOVERNMENT AND FOUNDATION GRANTSMANSHIP

"The real challenge to human ingenuity, and to science, lies in the century to come. I cannot guess at the things we will need to know from science to get through the time ahead, but I am willing to make one prediction about the method: we will not be able to call the shots in advance. We cannot say to ourselves, we need this or that sort of technology, therefore we should be doing this or that sort of science. It does not work that way. We will have to rely, as we have in the past, on science in general, and on basic, undifferentiated science at that, for the new insights that will open up the new opportunities for technological development. Science is useful, indispensable sometimes, but whenever it moves forward it does so by producing a surprise; you cannot specify the surprise you'd like. Technology should be watched closely, monitored, criticized, even voted in or out by the electorate, but science itself must be given its head if we want it to work."

Lewis Thomas
*Late Night Thoughts on
Listening to Mahler's
Ninth Symphony*

"Money-getters are the benefactors of our race."

P.T. Barnum

At this moment all over the country, countless professors, scientists, engineers, and project managers are no doubt reducing fossil fuel supplies by burning midnight oil to write proposals for government and foundation grants. On some campuses grantsmanship rivals team sports as the number one competition. The reason for this activity is simple. Government agencies and foundations have the money and sometimes the inclination to give science its head and to fund the sort of scientific activities that produce the surprises Dr. Thomas refers to as the challenge of the future.

Indignant critics sound off about researchers selling their science to the government or to a large foundation which through beneficent tax provisions is really just another branch of government. But the midnight oil burners can ignore the critics as misinformed or idealistic purists. "Who else will finance this project if not the government or a foundation?" the scientist asks, and their accusers have no answers generally that will supply enough cash to pay for sixty seconds of computer time.

Critics of the government and foundation grant process argue that science sponsored in this manner isn't free or runs the risk of not being free. The word "free" is a complicated idea in this context. Was Columbus free when he accepted the patronage of Isabella to find a new route to India? He didn't find one, but the Queen didn't order his head removed. The 500th anniversary of the event was being commemorated in various ways worldwide as this book was prepared, which demonstrates the long-term impact of some government-supported projects.

Government agencies and foundations generally do not make much of a fuss when the R&D projects they sponsored don't deliver all of the anticipated goods. Such grants at least seem to tolerate the right to fail, a right that private sector sponsorship of research is prone to deny. When a ball team loses games, the manager goes. In private industry research, when a project doesn't pay off, the manager may go or be transferred. Government and foundation grants show more tolerance of failure as an inevitable side of the scientific life and in this sense do give science its head to pass through the trial and error stage leading to a new surprise.

The problem is more complicated than the simple one of possible but not proven government-foundation interference in the scientist's unfettered need to think and do his work. Another face to this coin is the probability that scientists will voluntarily compromise or limit their work in order to win those all-important grants without which they may not be able to do the work at all. W.I.B. Beveridge made this comment about the situation in *Seeds of Discovery*:

> *"Under the grant system, researchers at universities or research institutes propose a research project and apply for funds from a grant-giving body, which may be financed from government funds or a charitable foundation. It has become accepted that the successful fund-raiser must devote much time, usually some weeks, to work-*

ing out his programme. It is commonly known that some researchers are more clever and successful at drafting applications than others; but 'grantsmanship' is a poor indication of true merits. There is a strong temptation to exaggerate the importance of the proposed investigation and the likelihood of obtaining valuable results. Applicants who are modest about their expectations and who refrain from committing themselves to a line of action for the following one to three years because they honestly cannot forecast it, are not looked on favourably by administrators and committeemen who sit in judgment. Many scientists have become cynical and say if you want to get the funds you have to play the game according to the rules laid down by those controlling the money."

So what else, we wonder, is new. The research game has pretty much always been played that way, with the possible exception of the era when Benjamin Franklin was free to go fly his kite. As long as money is necessary for scientists to do science, they will continue to seek it where it is available. The facts of research in the 1990s and beyond to the foreseeable horizons of time are that government agencies and foundations will remain important sources of R&D funding and perhaps the only source for projects that are not rich with the promise of early commercialization. Private corporations may fund such research through their corporate foundations, and some may put considerable money into long-term, speculative science. But corporations will not fill much of the need in this area, because they feel an impulse to show profits or they develop chronic red ink indigestion and their stockholders make groaning noises. Venture capital sources can never rival government and foundation sources as generous funders of projects that clearly will not double or even return the funder's investment in a short time.

There is no particular cause for astonishment or alarm that professors and many others work overtime preparing proposals to go after the available government and foundation grants. They obey nature's law of going or trying to go where the action (*i.e.*, money) is. Richard Mitchell in his book on the sad state of American literacy, *Less Than Words Can Say*, said this:

"In the whole United States of America at this moment there are only twenty-seven professors who have not at one time or another sat around with a few accomplices and plotted to think up a good gimmick for a government grant, and the wording of the prayer [proposal] that would bring the money. (A grant from some foundation is even better; those G-men are fussy about little things like bookkeeping and expense accounts.) Grant-getting has become one of the regular duties of some professors and administrators. Many colleges and universities have established whole bureaucracies devoted to grant-getting."

Some institutions and individuals are highly successful at finding, pursuing, and winning grants. The dilemma for most, however, is not that they have to do it, but that they don't do it well enough. Three common problems frustrate their efforts to win more grants from Squire Philanthropist and Lady Bountiful, alias government agencies and foundations.

1. Projects that aren't original. A routine comment by funding sources is that many proposals offer them what has already been done. "Don't plow a field that's already been plowed," said a Department of Defense program director. This problem is solved by making a thorough literature search, by knowing what has been done, by learning the specific needs of the agency or foundation.

2. Failure to follow instructions. There are two solutions to this problem that are amazingly common. The first solution is to read and the second is to follow the instructions.

3. Poor proposals. The solution is not as easy as reading instructions, but there is one: Write better proposals.

Repeatedly proposals have failed not because the project was a bad one or the scientific team unqualified but because the proposal simply couldn't begin to make an adequate case for either the project or the team. With such a proposal, Columbus would still be warming a chair in Isabella's outer office. Concerning the modern government-foundation situation, a program director made the following observation:

"The strange thing is that some of them won't let you help them. They borrow an out-of-date guideline brochure from a college down the road, dig up the wrong application form somewhere, and mail in a proposal just under the deadline. There is no time for discussion or revisions. I have on occasion telephoned such an applicant and suggested that we send him the current guidelines with the appropriate forms and that he rewrite the proposal for the next competition. Instead of being grateful for the help, he thinks he is being given the brush-off and insists on his right to have the proposal considered in the current round. What can you do? It's like watching a sinking ship—all you can do is stand there and salute."

Frustration obviously isn't limited to those who submit proposals. The reviewers and the grant makers are also subject to the same distress and concerns on the other side. The source of grants, whether government or foundation, must make grants to do its job; but how can those responsible reach informed decisions if the proposals they see are incomplete, badly organized, inconsistent, careless, and sometimes preposterously inadequate as documents intended to provide specific information about a project and the people who want to do the project.

A 1989 newsletter, *InKnowVation*, described a proposal as "a working, communications document which must basically provide the information needed by an evaluator to consider four conditions: *What* work is to be done—*When* it is to be done—*How* it is to be done—*Who* will do it." A dismaying number of proposals neglect to cover What, When, How, and Who. So why surprise when they produce a resounding thud?

Government and foundation grant sources have also in some cases been made wary by the aggressive and even unethical tactics used by some to gain their attention and win their grants. Virginia White in *Grants* commented on the rationalization by various grant-go-getters that anything goes "if the cause is noble, the proposal sound, and the applicant eminently qualified and sincere in his intent." These judgments, of course, are in the eyes of the proposer. The result of some corner cutting practices by a few grant seekers has been to complicate the whole process of getting grants in various programs and from some sources. "Granting organizations have all they can do to discriminate between the merits of the proposals submitted to them without having to simultaneously fortify themselves against the techniques of their presentation and advocacy," wrote Virginia White.

Competitive programs such as the federal Small Business Innovation Research (SBIR) Program are well-fortified against departures from the rules. Even a minor infraction as innocent as going slightly over the 25-page length limit will cause a proposal to be rejected. Such rigidity has the positive feature of assuring all applicants that their proposals will be identically processed and evaluated. Each department participating in the SBIR competition has established procedures guaranteeing such objectivity and fairness.

Proposal Do's and Don'ts

The government agencies and foundations that sponsor grant programs are normally very accessible and generous with information on proposal requirements. Full advantage of whatever hospitality is offered should be taken by anyone expecting to submit a proposal in a competition. What you don't know about the program and the sponsor can hurt you critically. Everything you learn becomes fuel for a better proposal.

At various conferences on R&D opportunities, SBIR program managers discussed proposal preparation and made suggestions based on their experiences in earlier competitions. The suggestions pertain to the SBIR Program, but they are also relevant to other grant programs and deserve attention.

The aims of the SBIR Program were summarized as follows:

To foster technological innovation in the United States.

To encourage and assist small firms in developing technological innovations by participating in government research.

To promote the transition of innovative research and technological developments from the demonstration phase to the commercial application phase.

Federal representatives with responsibilities in the SBIR Program repeatedly emphasized the following points:

1. *The Main Rule*: Follow instructions and use your common sense.

2. *Please read the solicitation*: Each agency publishes its own solicitation. The same guidelines are followed in the SBIR Program, but agencies run their programs differently with varying due dates, review methods, formats, etc. You should read carefully the specific solicitations you plan to act on.

3. *Use specified forms and formats*: The exact requirements of the agency receiving your proposal must be met.

4. *Address one topic per proposal*: The general rule is "One topic (or subtopic) — one proposal."

5. *Get the help you need*: Obtain expert advice from universities or consultants to strengthen your project and proposal, thus saving time and effort.

Each agency cited the common horror story — proposals that were not looked at because of errors resulting from neglect in complying with instructions. The wide-ranging comments tended to repeat or echo shared concerns.

Department of Agriculture

"Proposal reviewers give double value to scientific and technical merit. Innovativeness and originality receive special emphasis."

"Those who dominate the competition and win repeatedly are those who invest the time and energy required to prepare sound scientific proposals."

"The probability of commercial success is an important criterion in the choice of proposals for funding."

Department of Defense

The representatives stressed the importance in this large department of sending proposals to the right place and obeying instructions. "Unless your proposals reach where they're supposed to go within the department, failure is inevitable," one advised. He also emphasized the necessity of a careful literature and database search to learn what has been done, thereby avoiding repetition of work and assuring that proposals cover new ideas and approaches.

This statement parallels a similar observation by the National Science Foundation representative who said, "Examine in detail the seminal articles in your field and make certain you know what is going on at the forefront of your science."

Specific proposal tips made by Department of Defense representatives included:

"Many proposals we receive are too vague, and sometimes they violate the laws of physics."

"Proposals may be technically sound but have no application to our needs."

"A good summary is a positive feature in a proposal."

"Don't offer us something already done or in process. We want innovative ideas and solutions, not a rehash."

"Stick to the topic. Don't address multiple topics with one proposal."

Department of Education

"Debriefing is beneficial if you want to try again. It alerts you to faults that can be eliminated in the next competition round."

"Failing to follow instructions caused 14 of 200 proposals to be ineligible for review."

"The SBIR 25-page limit is absolute, 26 pages and you're out. That happens all too often."

"Studying the abstracts of winning proposals is a profitable tactic to help you succeed."

"Some of the specific factors that caused proposal rejection are these:
— Proposal does not give cost benefits.
— Proposal does not discuss the need for the product.
— The strategy for demonstrating feasibility is unclear.
— The plan is not original or innovative.
— The amount of work outlined and the estimated number of hours show a lack of realism.
— The evidence that those preparing the proposal did their homework is not apparent."

Department of Energy

"Always start with a good project idea."

"Keep up-to-date on current literature."

"Have an expert recognized and published in the field review your proposal criti-cally."

"Prepare the proposal carefully, follow solicitation directions exactly. Revise in accordance with the reactions of experts and other reviewers of preliminary drafts."

"Consider asking a university authority to join you as a consultant on the project and include that person in your proposal."

"Mail on time. Late is out."

The point is made again and again concerning grant competitions that the same rules apply to all and must be obeyed, such as delivery by the due date and with no more than the maximum number of pages. The persistent message is not to try even a slight deviation from the instructions, assuming an exception will be made in your case. It probably won't be made for you or for anyone else, which should be some consolation.

Department of Health and Human Services

"Never assume the reviewer of a proposal will know intuitively what you mean. Explain carefully, fully, and clearly."

"Conduct a thorough study of all related publications. Is your project really new?"

"Label diagrams and tables adequately so the reviewers can understand them without struggling."

"State the rationale of your project as specifically and comprehensively as you can."

"Run the proposal through as many drafts as necessary to make the language smooth and the organization logical and consistent."

"Just skimming the instructions won't suffice."

"The most frequent reason for proposal failure is the lack of an original idea and a true innovation."

Department of Transportation

"Excellent ideas can be defeated by careless mistakes."

"Inherent flaws due to ignoring requirements stated in the solicitation resulted in 37 of 412 proposals failing to pass administrative muster."

"Too many proposals must be rejected administratively before they reach the stage of technical evaluation simply as a result of not following the rules."

"Information engineering is a new and growing area of emphasis for the Department of Transportation with the potential for tremendous future impact. Traditionally

we've emphasized the mechanical and civil engineering side of transportation. Now we're in the period of information science and engineering."

Environmental Protection Agency

"To succeed in SBIR, firms should undertake cutting edge, high-risk or long-term research that has a high potential payoff. Continuing environmental research is essential to keep the flow of knowledge from slowing to a stop."

"The same research often can serve as the basis for technological innovation and new products or services that benefit the public. Such proposals are encouraged."

"An excellent proposal that is complete, explicit, and scientifically sound is a prerequisite for success."

NASA

The representative from the National Aeronautics and Space Administration said, "Remember we are typically deluged with nearly 1,000 proposals. We can't read your mind, all we can read is your proposal. Don't let carelessness or a minor administrative neglect cause you to lose. Call us if there's something you don't understand in the solicitation."

He mentioned that one proposal in ten succeeded. This low percentage of success occurred primarily because barely half of the proposals received offered something new, and many proposals that might have involved true innovations showed no evidence that the idea had been worked on sufficiently to establish a bona fide project with achievable objectives. The NASA representative specifically advised:

"Make the proposal fit one of our subtopics. Without a close match between the project and our subtopic, the proposal can't succeed."

"Show clearly that the project will lead to a measurable outcome. What will we have after Phase I and Phase II? Spell it out."

"Use a checklist to be certain nothing vital is omitted. Be realistic, make sense, and answer one of our particular needs."

"A proposal may be well written and yet fail to identify the innovation involved so that reviewers grasp the essentials. Make certain the proposal gets across the key points so they can't be missed."

The NASA representative made this request to those submitting proposals to NASA: "Please put yourself in our place and help us understand. In the proposal identify clearly what your idea is. Give a detailed explanation that will persuade a knowledgeable person your idea will work and that your goals can be reached." He added a reminder that if a proposal does not suit one agency, the proposal is not for that reason necessarily finished. It may be submitted to other agencies for consideration.

National Science Foundation

The SBIR program manager for the National Science Foundation (NSF) noted that "dead-end research" is not the goal of the program. At NSF the effort in the program is to fund research that demonstrates highest quality science and engineering and has strong commercial potential. And apparently NSF has been able to achieve the goal of quality research that pays off. NSF pioneered the SBIR concept with the first Small Business Innovation Research Program in 1977. This NSF program became the model for the SBIR Program put into effect for the entire federal government with 1982 legislation. Consequently, NSF has many years of experience with the program. Based on long experience reviewing R&D proposals, the National Science Foundation published a list of "Common Proposal Problems" encountered by NSF and other government agencies in R&D proposals. Following is a paraphrased summary of these "Common Proposal Problems."

Common Problems in R&D Proposals

Low Quality Proposals: A poor proposal shows inadequate knowledge of the subject and relevant research. It contrasts with a careful, well-organized proposal that persuades reviewers the project is important and that the firm submitting the proposal can do the work skillfully. The proposal should be prepared with the same attention to accuracy and technical detail as a paper for a refereed journal. The proposal should be self-contained and complete in itself. A good bibliography is a convincing component in a quality proposal.

Proposal Balance: Irrelevant material in a proposal weakens its impact and distracts from the main points. All proposal contents should be directly relevant, and the proposal should mainly focus on project objectives, the work plan, technical problems, and their handling. The proposal should include details of the proposed project, the qualifications of key personnel, and discussion of related R&D projects carried out by the applicant. Diversionary or sideline information should not appear to confuse reviewers and fog the message of the proposal.

Not Following Instructions: Often proposals show little effort to comply with the instructions given in the RFP or solicitation for a grant competition. This neglect virtually assures failure. Essential for success is using the prescribed format and supplying all requested information in an orderly manner that reviewers can readily follow. Reviewers consider only what they find in the proposal before them, and they require certain contents and arrangement of components. Why jeopardize your proposal's chances by ignoring rules that are issued for a reason and that are no trouble to follow if you learn what they are.

Allowing Too Little Time: Haste makes mistakes. Trying to prepare a proposal in too short a time can show the familiar symptoms of hurry—bad editing, clumsy writing, bad proofreading, inconsistencies, errors. Start work on the proposal well ahead of the date it is due and devote whatever time you must to the job to do it correctly.

Carelessness: If you are casual about the proposal, careless about its writing and contents, you should assume that many competitors will treat the challenge very seriously. If you submit a proposal demonstrating carelessness, you in fact surrender to the competition in defeat. You should not allow your work to be represented by a weak proposal that gives a negligent impression.

The National Science Foundation in connection with small business proposals for R&D grants made the following comment in its advice on proposal preparation:

> *"One approach small businesses can take . . . is to seek the review of the proposal by a university-based scientist or engineer who is highly knowledgeable on the subject and who is involved in research or R&D. Many may be interested in reviewing proposals directly in their area of interest, particularly if they are to be included as a consultant if the proposal is funded. University Small Business Development Centers, state university engineering extension services, and sponsored program offices sometimes assist small firms in finding such potential consultants. It might be added, that university consultants often strengthen a proposal in the minds of reviewers as well as in their working with the small business."*

Two main axioms emerge from the comments by the SBIR Program representatives: *first*, that the quality of the science involved and the validity of the project are the chief criteria in proposal evaluation, *second*, that the instructions mean what they say and you'd better obey them.

Though intended for the SBIR Program, the suggestions about proposals by federal officials reported above have wider pertinence for most grant competitions and proposal preparation in general. Based on the views of those who work at the front lines where proposals are received and evaluated, we can list the following recommendations as ways to produce more convincing proposals and as a result to qualify for more government and foundation grants:

1. Give yourself an up-to-date background by keeping current on technical literature.

2. Contact published researchers to review your proposal and to be listed in the project as consultants. The names of recognized authorities strengthen and validate a proposal.

3. Seek interviews with grant winners at universities, research institutes, and elsewhere. Learn from their experience.

4. Forge links with university researchers if you operate a small business. Such collaboration often stimulates the R&D process for both parties.

5. Study the government agency or foundation which will receive your proposal. Identify their specific activities/programs/needs and respond accordingly.

6. Get your name on the Small Business Administration mailing list and other government mailing lists. Make personal contacts at the offices you hope to serve. Take any steps that will keep you informed on government RFPs. When an RFP matches your skills and interests, write a proposal and get it in for consideration.

7. Try hard for the grants of your choice. Enter the competition to win. If a proposal loses at one agency, send it to another that you check out and think may be interested. Try the same agency again where the proposal failed with a new proposal (perhaps the old one revised) when they issue another RFP. Grant winners are those who persist despite rejection.

8. Consider submitting unsolicited proposals to the agencies that encourage them. Check out the agency that seems a likely funding source for you, and give it a try.

9. Master agency and foundation requirements for proposals and learn the details of their various grant programs. The more you learn about them, the easier it will be to prepare proposals that meet their special needs.

10. Make your capabilities known to agencies and foundations. Vigorously go after the government business and the government or foundation grants for which you are qualified.

The billion dollar plus opportunity for R&D grants offered by government and foundation programs makes proposal preparation aimed at these targets highly worthwhile if you know realistically that your project has a chance. In fact, ignoring the opportunity if R&D funding is needed to carry out a project with great prospects would be poor tactics. Geese that lay golden eggs and treasure pots at the ends of rainbows are a lot less plentiful than we might wish. But there are government and foundation grants. Someone will qualify for each grant with a proposal that satisfies requirements. Why not your proposal?

Finding Out About Grants

You can't prepare a proposal and try for a grant unless you know about the opportunity sufficiently in advance to take action.

Various tips are given above and in Chapter 4 regarding ways to keep track of government and foundation grant programs. The following list summarizes things you can do:

(!) Get your name on government and foundation mailing lists. Read the solicitations and announcements they distribute.

(!) Subscribe to *Commerce Business Daily*. This puts you in touch on a continuing basis with government business opportunities.

(!) Contact the government agency or foundation that interests you directly for information and guidance about research needs and proposal requirements. Telephone or write grant program officers for firsthand assistance.

(!) Study publications of the Foundation Center and follow up the leads you acquire.

(!) Attend conferences on R&D opportunities sponsored by state and federal groups.

(!) Communicate with researchers, managers, and grant winners in your field on a regular basis to keep updated about where the action is.

(!) Fully utilize the resources in science and business libraries at universities or colleges and in the reference departments of good public libraries. Libraries can sometimes surprise you with the scope of their support materials.

Databases

Databases as valuable research and support tools are discussed in the "Computers and Prewriting" section of Chapter 6. A large number of databases are available to help you acquire vast amounts of information relevant to this inquiry and to obtain technical support data for your research project.

A vast number of books and magazines, with the total growing constantly, give current details on commercial and government databases covering most categories of modern scientific, technological, business, and social concern. Your public library probably has many of these information resources.

Government databases in particular should be checked out by those intending to submit proposals for government and foundation R&D grants. These databases can give invaluable information concerning what has been published on your subject and much more data to back up your efforts. Government databases are accessible free or at a nominal charge and provide information resources second to none if your goal is to submit winning proposals for federal R&D grants.

Many of the departments and agencies that offer SBIR grant programs have databases that can be accessed by researchers with established eligibility. Learning what the requirements are to take advantage of these services and following up is recommended procedure for those who want to do business with government organizations. The diversity of government databases is tremendous—for instance the Department of Interior has a Mourning Dove Database as an index to breeding trends and hunting seasons. The Defense Technical Information Center (DTIC) in Alexandria, Virginia is operated by the Department of Defense to assist anyone doing business with the military, such as grantees in the SBIR Program. The DTIC has several million records, and the collection is continually updated. Using DTIC services can protect applicants against submitting proposals on research that has already been done.

Contact: Defense Technical Information Center
ATTN: DTIC-SBIR
Building 5, Cameron Station
Alexandria, Virginia 22304–6145
Telephone: 202-274-6902; 800-368-5211 (Toll Free)

Many federal and some state government sources supply free information and even perform research in various areas of business, energy, and technology.

Those who prepare proposals should find out what these services are and where they are. Often if the initial contact point cannot supply the information needed, you can be directed to databases or other sources that can help meet the particular need.

Government databases and information services can frequently help you with time-saving, cost-saving data. Since taxes subsidize these services, you should get your share of their basic product — information.

Resources and Contacts

If you have trouble locating the right expert or group of experts, contact the nearest Federal Information Center for initial guidance.

The SBIR Program Solicitations issued by the participating federal agencies (free on request) typically include detailed sections with Scientific and Technical Information Sources available to assist you in preparing strong, well-documented proposals. These sources constitute a veritable Ali Baba's cave of useful information. Obtain the SBIR Solicitations for the departments or agencies that interest you and follow up these vast resources.

A Department of Energy (DOE) SBIR Solicitation, for example, describes the DOE Office of Scientific and Technical Information (OSTI) which "collects and disseminates both DOE-originated and worldwide scientific and technical literature in subjects of interest to DOE researchers. This information is stored in a comprehensive group of databases on energy and is available to other government agencies, universities, and the general public in varied media." The DOE Solicitation indicates that the Energy Database with over two million scientific and technical references is available, and OSTI also operates an on-line Superconductivity Information System which includes a bulletin board and three databases.

Comparable data and information resources are maintained by most other government departments and agencies. Check them out.

Contacts you should know about include the following among many:

National Technical Information Service (NTIS): Reports resulting from federal research, and reports received from exchange agreements with foreign countries and international agencies are available through NTIS for a nominal fee.

U.S. Depository Libraries: The Government Printing Office maintains depository collections of extensive, unclassified government materials at hundreds of public, college, and university libraries around the nation. These are known as U.S. Depository Libraries. Check with your public library, which may either serve as such a Depository or be able to help you contact the one nearest to you.

Library of Congress: Invaluable technical reference and bibliographic services, both free and for a fee, are offered by the Library of Congress.

Library of Congress Switchboard Telephone: 202-707-5000

Science and Technology Division
Library of Congress
10 First Street, S.E.
Washington, D.C. 20540
Telephone: 202-707-5664

This Library of Congress address and the telephone numbers came from the 1990-1991 edition of the *Washington Information Directory*. The Directory is published annually by the Congressional Quarterly, Inc., 1414 22nd Street, NW, Washington, D.C. 20037 [Book Division: 300 Raritan Center Parkway, P.O. Box 7816, Edison, N.J. 08818-7816] and is a useful single source on departments and agencies of the federal government, Congressional committees, and private nonprofit organizations that are Washington-based. The Directory is available at many libraries. It also can be purchased by those whose volume of Washington contacts warrants the book's constant availability.

Washington Information Directory contacts for information and orders: Telephone Toll Free: 1-800-543-7793
FAX for orders: 201-417-0482.

The local or Washington office of your Senator or Congressman is another useful place to start the process of identifying, contacting, cultivating, and productively exploiting the particular information resources that can serve you best. Also, don't neglect to make full use in this context and many others of an indispensable resource for proposal writers, the public library.

Foundation Center Databases

Among numerous databases that can assist you in the private sector are those available through the Foundation Center, which is described in chapter 4. In addition to publications that will concern many who prepare proposals, the Foundation Center is also a database resource. The 1991 edition of *The Foundation Directory* states, "Perhaps the most flexible way to take advantage of the Foundation Center's vast resources, computer access lets you design your own search. . .On-line retrieval provides vital information on funding sources, philanthropic giving, grant applications guidelines, and the financial status of foundations."

Foundation Center data are available on-line through DIALOG Information Services and through many on-line utilities. Information about accessing the Center's databases through DIALOG is available from DIALOG at 415–858–2700. The On-Line Support Staff at the Foundation Center is available at 212–620–4230.

Saying Yes to Databases — Helpful and Informative
Friends of Well-Based Proposals

Now that we have entered the age of computers talking to computers, knowing about and using the computerized databases available in your field are becoming almost compulsory to keep up and compete. Over a quarter of a century ago, business professor Thomas L. Whisler presciently predicted a world "where the organization will come close to consisting of all chiefs and one Indian. The Indian, of course, is the computer." This colossally effective, informed, speedy, and versatile "Indian" for proposal preparers in need of support data is the computerized database. By supplying helpful information quickly, databases can be important resources when you prepare a proposal. Go introduce yourself to databases. Step up and say hello. Learn about them, find out how to access them, how to use them. Your dividend from the effort will be stronger proposals.

Taking the Grant Gamble

Developing a project and preparing a proposal to compete in a grant competition may seem at times like a form of scientific roulette. A great effort is needed even to lose, and there are usually many more losers than winners.

So why bother? You bother because the opportunity is genuine and the stakes are real. And not everybody loses. There really are winners too. Chances are you would try somehow to carry out the project even without a federal or foundation grant. So why not go for the grant. You have nothing to lose except the time, trouble, expense, and hard work of preparing a proposal—and even that may have other uses as a marketing or planning tool.

Also include the fact that you have quite a bit of faith in your project, and the doors won't open unless you knock. Such arguments convince many to try, hoping that if they try hard enough their project and proposal will make it to the winner's list. This leads to the basic rules of grantsmanship:

1. Don't lose by not trying.

2. Write your proposal.

3. Submit it.

4. The results may delight you.

CHAPTER 11

VENTURE CAPITAL PROPOSALS

"Nought venture nought have."
John Heywood
Proverbes, 1546

"There are one hundred men seeking security to one able man who is willing to risk his fortune. . .I buy when other people are selling."
J. Paul Getty

The idea that one must take a chance to win is a very old human notion as shown by those ancient nuggets of folk wisdom called proverbs. "Nothing ventured nothing gained" is several centuries older than capital gains, but the proverb is probably not as old as the fact of venture capital. The history of research and development is packed with success stories in which a timely investment of venture capital launched a major winner from Noah's Ark to the most famous names in modern business—and often paid off handsomely for the investor.

Venture capital firms and venture capital investment divisions of some banking institutions today control billions of dollars, and all of them are on the lookout for highly promising risks. Start-up companies with radically innovative and unproven products or ideas—such as photocopying, Polaroid, transistors, videotape recorders, home computers, microprocessors, compact disks, high-definition television, etc.—could find they have no possible source of financing except venture capital, risk money seeking a winner.

The challenge to the business or the individual with a bold new notion is convincing a venture capital firm that the idea has the potential to make it big and assure the high, quick rate of return venture capital organizations require. A venture capital project must show considerable

125

potential and realistically hold out the golden apple of profitability within 3–5 years and tripling the investment in 5–7 years.

How do you make contact with a venture capital firm and inaugurate the process of establishing a relationship that may finance your speculative enterprise?

The answer: A proposal.

A venture capital proposal includes many of the elements found in the formal, elaborately structured proposals submitted in government and foundation grant competitions. There are differences, however, in handling and presentation that should be recognized.

"There isn't a set of rules imposed from the outside that we have to follow. That allows a more creative process," said a venture capital firm executive. He emphasized that what he needs are the facts in a compact, manageable format. Referring to a massive proposal he received with more packaging than substance, he said, "The proposal lacked the basic things I had to know. It didn't answer my questions. That kind of hype doesn't travel well with me or other venture capitalists."

Among the criteria that venture capital firms consider when making equity investments for partial ownership interest in a firm are:

Experienced Management Team

The venture capital firm likes to find a strong management team in place with expert knowledge of the field and proven executive strengths in management, marketing, finance, and production. Ideally the team has a track record showing the different members work well together for creative, dynamic results.

Unique, Innovative Products/Services

The products are usually past the R&D stage and into prototypes or actual production. The products or services should be proprietary, unique, with distinctive sales appeal. The venture capitalist likes exclusive products that are or can be patented and give a company important lead time in developing a market.

Potential for a Large Market

The potential market should be half a billion or more and there should be a chance to expand dramatically in less than a decade.

When your situation reflects these criteria or can in time and with venture capital assistance achieve them, you are ready to seek venture capital.

A U.S. Small Business Administration publication on venture capital, "Management Aids for Small Manufacturers," (MA 235) indicated that venture capital firms typically seek investments of $250,000 to $1,500,000. "Projects requiring under $250,000 are of limited interest because of the high cost of investigation and administration; however, some venture firms will consider smaller proposals if the investment is intriguing enough," stated the management aid. The publication noted that most venture capital firms receive over 1,000 proposals a year each and that "Probably 90 percent of these will be rejected quickly because they do not fit the established geographic, technical, or market area policies of the firm—or *because they have been poorly prepared.*"

Example 11 reproduces "Elements of a Venture Proposal" from this publication. Small Business Administration management publications are worth consulting for good ideas to use in financial planning. Small Business Administration publications and services are available through district offices or the main office in Washington, D.C. (see Chapter 4 for the address).

The Business Plan

An alternative to the proposal that is often used for several purposes, including raising capital from investors, is the business plan. Some venture capitalists like to start learning about an organization by reviewing the business plan. A standard way to approach a venture capital firm is to send the firm a business plan with a covering letter that discusses the financial assistance sought in the form of venture capital. In effect the business plan and covering letter plus any amplifying documents added to the package constitute the elements of a venture capital proposal.

Among the most important steps when starting a business is developing a business plan that identifies business objectives, needs, opportunities, and problems. The business plan serves a number of essential functions including:

1. Assists in determining the feasibility of the new business and its prospects for success.

2. Provides the basic information for development of a business operational plan.

3. Indicates the steps necessary to start the business.

4. Serves as a document to raise capital from venture capitalists or other investors.

Preparing a business plan closely resembles the preparation of a proposal, and they have many features in common. A proposal essentially is a business plan for a single project rather than a business as a whole. Taking special pains to organize a strong business plan helps a business avoid difficulties, solve problems, and expedite progress.

Example 11: Elements of a Venture Proposal

Source: *Small Business Guide to Federal R&D Funding Opportunities* (March 1983, PB83-192401), prepared for National Science Foundation. Reprinted from MA 235, "Management Aids for Small Manufacturers," U.S. Small Business Administration.

Purpose and Objectives—a summary of the what and why of the project.

Proposed Financing—the amount of money you will need from the beginning to the maturity of the project proposed, how the proceeds will be used, how you plan to structure the financing, and why the amount designated is required.

Marketing—a description of the market segment you've got now or plan to get, the competition, the characteristics of the market, and your plans (with costs) for getting or holding the market segment you're aiming at.

History of the Firm—a summary of significant financial and organizational milestones, description of employees and employee relations, explanations of banking relationships, recounting of major services or products your firm has offered during its existence, and the like.

Description of the Product or Service—a full description of the product (process) or service offered by the firm and the costs associated with it in detail.

Financial Statements—both for the past few years and pro forma projections (balance sheets, income statements, and cash flows) for the next 3-5 years, showing the effect anticipated if the project is undertaken and if the financing is secured (this should include an analysis of key variables affecting financial performance, showing what could happen if the projected level of revenue is not attained).

Capitalization—a list of shareholders, how much is invested to date, and in what form (equity/debt).

Biographic Sketches—the work histories and qualifications of key owners/employees.

Principal Suppliers and Customers: Problems Anticipated and Other Pertinent Information—a candid discussion of any contingent liabilities, pending litigation, tax or patent difficulties, and any other contingencies that might affect the project you're proposing.

Advantages—a discussion of what's special about your product, service, marketing plans, or channels that gives your project unique leverage.

A good business plan can be prepared in different formats. A business plan intended to sell an idea or a company to a venture capital firm should allow the reviewer of the plan to learn quickly the answers to these key questions:

(?) Who are the present or potential customers. What's the size of the potential market.

(?) Why should a customer choose your product rather than buy from your competitors.

(?) What are the features and benefits of your product.

(?) What technical issues and problems remain to be solved.

(?) What are the realistic future prospects.

Business plans vary depending on the nature and needs of a business. Whatever elements a plan includes, it needs to deal objectively with problems as well as strengths and provide realistic information for reviewers and management.

Venture capital firms prefer business plans that are no longer than they have to be (this echoes a familiar goal with proposals). A business plan that is unnecessarily long and wordy for the information contained makes a weak impression. One venture capital firm representative advised, "Keep your business plan under 15 pages, and the entire plan will be read. If it is over 15 pages, include a summary for those people who don't read long reports. That's the way to make certain you have a real chance to put your ideas across."

When advice is given on how to improve business plans, the comment from venture capital firms most often heard is to strengthen the marketing discussion — more information about existing and potential markets, how markets will be developed and why the marketing plan should work.

The following business plan outline (Example 12) lists information that might and frequently does appear in a business plan. An individual business should adapt the outline to its special circumstances. The detailed business assessment required to prepare such a business plan will help management grasp the current state of the business and make informed decisions for the future of the business.

Since finding a source of venture capital may be the only quick way to obtain crucial financing for a business or project — waiting eight months for a grant might kill the deal — learning how to meet the requirements of venture capital organizations is good business and also good science if the effort helps you turn an idea into a future winner. A venture capital partner can be an asset for small companies getting started by supplying strengths where weaknesses exist, such as professional management and

Example 12: Business Plan Outline

Cover Sheet: Name of business, principals, and location.

Statement of Purpose: Company objectives, what your business is, essential elements of your business plan, your overall business concept.

Business History: Past and current status of the business, historical development, description of present product or service lines.

Product or Service: Discussion of your competitive advantages at present and projected.

Market and Industry Analysis: Description of the industry, its major characteristics, current status, and trends. Discussion of the market including size, segments, competition, customer outlook, prospective development.

Marketing Strategy: Present and proposed marketing activities, covering distribution, promotion, pricing, geographic penetration, and future priorities.

Management: Organizational chart, key individuals with biographical sketches and indication of their positions in the business, and planned staff additions or changes.

Financial Statements and Projections: Complete details on the financial status of the company including financial statements and projections for the next five years as well as historical financial statements. Indicate how the financial statements relate to industry performance.

Company Ownership: List holders of company stock and those with other types of equity in the company. Provide a clear profile of the company's ownership status.

Technology Status: Explain requirements, policies, and procedures in connection with production and operations. Describe the technical status of products, the patent or copyright position, regulatory requirements, potential risks and pitfalls, physical facilities, suppliers, the labor situation, the manufacturing process, new technologies ahead, the research and development program, and cost vs. volume curves with cost breakdown for each product.

Sensitivity Analysis: Best, worst, and most likely projections.

Funding Requirements: Estimate of the funds needed and the reasons for the need.

organizational assistance. When accepting venture capital support, how-ever, keep in mind that you will probably give up equity in the firm and some degree of control as well. The venture capital firm may leave oper-ating control to you but have provisions in the equity financing agree-ments that allow them to assume control and appoint new officers if circumstances require such steps to protect their investment. This should not be taken as a red flag against venture capital. It is simply one of many factors to weigh when you have an idea to protect and a dream to make real.

Whether or not you decide to seek venture capital, preparing a good business plan is an excellent tactic to learn precisely the current status of your business and to point out clearly where future directions lead.

CHAPTER 12

INTRACOMPANY PROPOSALS

*"There is no adequate defense, except stupid-
ity, against the impact of a new idea."*
Percy W. Bridgman
Nobel Laureate in Physics

*"I have received memos so swollen with man-
agement babble that they struck me as the
literary equivalent of assault with a deadly
weapon."*
Peter Baida, 1985

Much of the writing done by engineers, scientists, managers, supervi-
sors, and other employees concerns inside-the-organization, work-
related compositions including: Letters, Memoranda, Reports, Articles,
Speeches, Notices, Instructions, Guidelines, Brochures, Newsletters,
Warnings, Requests, Refusals, Documents, Visuals, Manuals, Leaflets,
Explanations, Handbooks, etc. . . .aimed at getting by, getting ahead,
and getting through the week in the work arena. Add proposals, of
course, to the long list of on-the-job writing media needed and used
regularly. The paper barrage in companies as well as governments supports the observation by Wernher von Braun that "we can lick gravity,
but sometimes the paperwork is overwhelming."

Communications at work reflect every form of human writing known,
from formal proposals to the CEO and the Board of Directors to an
unsigned graffito scrawled defiantly on a conspicuous wall when the
author hopes no one is looking, in order to deliver a strong message
anonymously.

The importance and scope of work-oriented writing have produced a
substantial literature on the subject aimed at both the classroom and the
individual. Two among numerous books that served both students and
the work force effectively are *How to Write for the World of Work* by

133

Thomas E. Pearsall and Donald H. Cunningham and *Successful Writing at Work* by Philip C. Kolin.

"To ensure a successful career, you must be able to write clearly about the facts, procedures, and problems of your job. Writing is a part of every job," Kolin stated in the Preface to the 1990 Third Edition of his book. "The higher you advance in an organization, the greater the amount of writing you will be doing." Intracompany proposals are likely to constitute a significant portion of this obligatory organizational writing.

Put It in Writing!

Ardent arguments are advanced in favor of across the desk, one-on-one communications whether inside or outside the company, with emphasis on the importance of direct human contact. Known as the BHS (Bunker Hill Strategy) rule, this modern managerial ploy comes from Colonel William Prescott's famous order to his steady band of Colonials, "Don't fire until you see the whites of their eyes" on June 17, 1775.

Following any one-on-one session — whether across the desk or across the continent via telephone — careful subordinates, wise managers, and conscientious communicators make certain they put in writing the gist of what was said and specifically note down all that was agreed on.

Transmitting these summary records to the other participants is optional but generally opportune in terms of career and company operational strategy. When it's in a written form that everyone involved has seen, any grounds for dispute are eliminated or at least clearly defined.

Trusting entirely to memories after decision-making meetings asks inefficiency to take charge and invites disagreement, argument, and trouble. That's why minutes are scrupulously prepared on all meetings in which several people are involved and why each participant in a two or three person meeting should take the trouble to prepare a memorandum about the discussion and the decisions reached. The procedure is practical, protective, and constructively supportive.

Exact words on paper continue to be the finest documentation and aide-memoire (memorandum) available. The most compelling, thorough, and incontrovertible form of inside-the-organization communication is the intracompany proposal.

The Valuable Uses of Proposals Within the Organization

Intracompany proposals are prepared for countless reasons and purposes. They can be short and sweet or intricate and elaborate depending on the cause.

Frank R. Smith in a discussion of "engineering proposals" offered a description that fits many intracompany situations: "In the conversation of business. . .a proposal is simply a formalized proposition. Written out in rather elaborate detail, the proposal merely says in effect, 'Here's what we will do for you at this time for this price.' In this sense it is no different from the proposition of the horse trader who says, 'I'll tell you what I'll do: I'll throw in the saddle and the bridle with the horse, just to make a deal.'"

A proposal is the vehicle you use to persuade others in the company to agree on and buy some particular "horse," notion, program, or project you have determined will benefit the organization in various ways that the proposal attempts to spell out convincingly.

All the rules, tips, and suggestions given earlier concerning effective proposals outside the company apply inside the company as well. But be prepared for complications resulting from office politics, knowledge of and interaction with colleagues, turf and position guarding impulses, and special personal considerations because you know the territory so well and are part of it. The success of intracompany proposals requires taking company political, social, and tactical facts of life and business into consideration as well as the inherent requirements of proposals.

Forms of Internal Proposals

Internal proposals may be of the solicited type. You are asked to make recommendations by your superior or by another department, and you do so impressively with a comprehensive, thought through, detailed action proposal.

Or the proposals may be of the unsolicited type. You see a company need or opportunity and submit a proposal via the proper channels for appropriate action.

Intracompany proposals come in a multitude of sizes and formats. Company proposals are not like gloves and hosiery; one size does not fit all situations. One form of simple and direct proposal is a note in the suggestion box about books in the cafeteria, formation of a company bowling team, or an easy way to improve the proposer's low morale by changing all that high-brow music on the speaker system to some civilized C&W, please!

Other intracompany proposals may be every bit as intricate and important to the employees collectively and to the entire company as major proposals to funding agencies in pursuit of R&D grants. Internal proposals can involve momentous decisions and large capital sums.

An internal proposal may work in the form of a memorandum, a letter, or a report. It may follow the full-scale proposal format with most or all of the elements examined in chapter 3, or it may utilize a modified format. Whatever format is chosen, the internal proposal should cover in detail the specifics about what is to be done, when, how, and by whom, with a cost breakdown and what the benefits will be to the organization from acting on the proposal. Here is an outline of a multi-purpose proposal that serves competently in many situations and is flexibly adaptable:

Example 13:   Intracompany Proposal Format

To:	Name of Proposal Recipient.
From:	Your Name and Title.
Subject:	Succinct Description of the Project.
Abstract:	Brief Summary (100–150 words).
Introduction:	Background statement on the need for the project and what will be accomplished.
Method:	Specific details about what will be done.
Work and Management Plan:	Cover the tasks that will be carried out [include a timetable], the facilities to be used [list specifically], names of the person(s) in charge with explicit responsibilities and duties, and budget figures for the entire project with the total costs.
Project Participants:	Names, qualifications, and functions of all the personnel involved.
Conclusion and Anticipated Results:	What will be achieved and the benefits resulting.

Situations Requiring Intracompany Proposals

When a new idea, recommendation, or project concerning any serious company issue is introduced by an employee with the objective of gaining its approval or adoption, a carefully considered and prepared intracompany proposal should be organized and submitted as the surest means of achieving success.

A proposal inside the organization has the identical purpose associated with any other proposal — to convince, persuade, obtain agreement, gain approval, receive permission and support to move ahead. The intracompany proposal that works is a convincing sales and information document for a plan, procedure, or action you want adopted.

The essence of such a proposal is not "soft sell" or "hard sell" but "effective and honest sell" with all the facts, figures, details, and perspectives needed to facilitate informed decision making. An internal proposal should reflect the company's best interest by frankly identifying all the negatives as well as the positives involved in the proposed activity.

The proposal may be submitted to a supervisor, department head, company executive, the President, or the Board of Directors, depending on the nature of the suggestion and the position in the organization of the employee.

The scope of the project determines whether the proposal is brief and uncomplicated or detailed and extensive. The purpose for which the proposal is prepared determines the extent and nature of the proposal document.

A one- or two-page simplified proposal-memorandum may suffice as a proposal to acquire a piece of equipment (e.g., copy machine, FAX machine, computer software) or to make a minor change that should increase efficiency and improve conditions. Here is a workable proposal-memorandum format:

Example 14: Proposal-Memorandum Format

To:	The Recipient and Decision Maker.
From:	Your Name and Title.
Subject:	Brief Description.
Purpose:	Reason for the Proposal.
Problem:	Why the Proposal is Needed.
Solution:	Recommendation for Solving the Problem.
Project Costs:	Accurate Budget.
Rationale:	Concluding Arguments for Proposed Actions and the Benefits.

This outline covers the essential elements for a short proposal presentation pertaining to a relatively uncomplicated situation or limited issue.

Generally, any inside-the-company proposal should 1) introduce the issue, 2) explain why it is a problem, 3) discuss your plan for solving the problem, 4) give the specifics of the plan including the costs, 5) conclude by supplying arguments for the plan, paying attention to the drawbacks if any, and pinpointing the benefits.

A proposal to revamp the company's operations in a substantive fashion, such as major departmental changes, extensively revised corporate policies, a new marketing system, etc., demands a more ambitious, well-substantiated document based on comprehensive facts and figures, carefully fortified with logical supporting arguments. Then a full-fledged proposal with multiple proposal elements and backup materials will probably be necessary to achieve full understanding of the project by all concerned and to obtain the necessary approvals required for concerted action.

Special Need for Sensitivity and Discretion

When preparing and submitting an internal proposal, be aware that it probably will affect and concern your fellow employees. You should realistically weigh and consider that fact in the interest of company harmony and project success. It is never smart to ignore office politics and the views of your colleagues when you make proposals whose success will require or at the least profit from their acquiescence and support.

Of course, intracompany proposals can also be issued as deliberate shake-them-up or wake-them-up declarations that you well know will produce upheavals. When you launch a bold campaign to stir and change in dramatic fashion and deliberately rock the boat, it is wise to realize in advance there are heavy seas and loud protests ahead as a result of your proposal. If that is the intention, you should be fully aware of the fact, choose your tactics accordingly, and be prepared for aggressive consequences. When you are part of a company, going it alone is not the best choice unless that is the only choice and you are fully cognizant of the situation.

In normal circumstances, you want as many as possible actively approving your proposal and on record to that effect. Conferring with the others involved before submission is highly advisable and often strategically shrewd. Obtaining the corroborating signatures of other employees on a proposed action that concerns them is always helpful.

Intracompany Proposals and Intrapreneuring

A special situation calling for sophisticated and detailed proposals is the growing trend in the U.S. for intrapreneurial spin-offs of small, independent operations within major corporations. These spin-offs sometimes emerge as autonomous new businesses. Typically they are the ultimate offspring of internal company proposals.

This occurs particularly when the large company has a new and promising technology outside its main line of operations that it chooses not to develop and market. Such a situation brings the loud knocking of opportunity for intrapreneurs. These are inside-the-organization equivalents of entrepreneurs who burn to turn neglected technology into products and profits.

When someone inside a company believes strongly in a new technology and hopes to develop its commercial possibilities, an internal company proposal is often the first step. If the intrapreneur wants to exploit the technology with the corporation's blessing, license to market, and even some of the corporation's capital to support the venture, the proposal must include detailed arguments and specific recommendations related to each objective, including clear indication of the benefits the corporation will receive.

Intrapreneuring Success

An intrapreneur at a major corporation recognized possibilities in a special computer chip that had been developed with his participation in connection with one of the corporation's R&D projects. When the corporation was not interested in the chip as a product, the intrapreneur learned about successful intrapreneurial models elsewhere and decided to emulate them at his organization.

"After studying the options, I decided the best course was to license the technology to a small business that could actively pursue other markets. Then came the challenge of convincing the corporation to do what it had never done before," he said. The intrapreneur, assisted by experts, went to work, devising strategy and putting together a proposal.

A proposal and business plan were prepared showing the corporation how it could benefit by licensing the technology and supporting the intrapreneurial development. The proposal and business plan needed were just as thorough as those required for venture capital or R&D grant applications.

The intrapreneurial enterprise was successfully started; and it made rapid progress because of the technology and because of the start-up firm's known connection to a corporation with global prestige and interests. A successful proposal prepared and submitted inside the company was the activating instrument for this enterprise.

Proposals Help You Help Your Company and Advance Your Career

Intracompany proposals assist you in implementing ideas and making the most of opportunities inside the organization. Taking steps to upgrade your written communications and to submit more detailed and effective internal proposals will pay in the long run. Carefully organized and submitted proposals will expedite obtaining approvals for your company projects, whether small or large; facilitate getting important tasks accomplished; and earn you a reputation for thoroughness, thoughtfulness, commitment, and efficiency. No kidding, such a reputation tends to benefit a career at least often enough to warrant the effort.

CHAPTER 13

PROPOSALS FOR ALTERNATIVE PURPOSES AND SPECIAL TARGETS

"Daring ideas are like chessmen moved forward; they may be beaten, but they may start a winning game."
Johann Wolfgang von Goethe

"This nation has said there are no dreams too large, no innovation unimaginable and no frontier beyond our reach."
John S. Herrington, 1987

A wide range of proposals for special purposes and particular groups deserves awareness and consideration. These are proposals well outside the new product development grant area and the R&D support realm. They do not necessarily concern innovative technologies but are important proposals nonetheless, because they too open portals of opportunity and seek support for meritorious activities and groups.

Among these proposals are those designed to interest book publishers and magazine editors with fresh ideas for books, articles, and columns. These are sometimes called "queries," but queries of this nature are legitimate proposals calling for care and skill in preparation to optimize the chances of success.

Another large category involves proposals from consultants offering expert services in specialized fields and specifically identifying how those services can benefit recipients.

Consultant Proposals

Wherever you find someone who has developed expertise in a particular field, you find a potential and probably an active consultant. A

141

consultant provides expert counsel and services to clients for a fee. To reach those clients, the consultant often must prepare and distribute a proposal. Ralph Waldo Emerson is associated with the view that if you build a better mousetrap, the world will make a beaten path to your door. Consultants often find it a little tougher than mousetraps. Even when a consultant offers a superior brand of expertise in a needed area, the world isn't likely to beat a path to the door. The consultant generally needs a competent proposal to inform potential clients about his availability and services.

Larry Greiner and Robert Metzger in a book on consulting suggested that an overview of a consultant's work proposal shows three segments: The *Beginning* explains why the proposal is submitted and describes the work to be done. The *Middle* covers the time required, the personnel involved, and the cost. The *Ending* discusses the anticipated results and the expected benefits to the client. These divisions encompass the traditional information found in most proposals.

Herman Holtz in a how-to book for independent consultants recommended a proposal format that stresses initially introducing yourself and your organization including details about your experience and accomplishments; discussing the particular problem you intend to solve for the client and the approach you propose; outlining the specific steps you will take and the staff, schedule, and cost details involved; and including references and other documentation with the proposal package as support materials to strengthen your offer.

These authorities and others on consulting emphasize the importance in consulting proposals of clearly demonstrating your experience and capabilities. The fact that "Ph.D." may appear after the consultant's name signals the acquisition of professional training in research, but equally important is clear-cut evidence of a successful track record in the field and effective service to satisfied clients. This information is an essential ingredient in a convincing proposal package from a consultant offering advice and services. Example 15 provides an adaptable format.

The consultant's proposal should include appropriate graphics—tastefully chosen to communicate and fortify, not to overwhelm. The proposal package can be strengthened with suitable publications by the consultant, references, brochures, and related items that complement the formal proposal without pushing it into the background. The proposal is the specific Action Element in the package. You don't want the client simply to admire and put aside your support materials. You want the client to respond promptly and favorably to your proposal.

Example 15: Consultant Proposal Format

NEED:	Discuss your perception of the need/problem and how you can provide a solution. Sell the idea that you understand the problem and are best equipped to deliver the answer.
PROCEDURE:	Explain what you will do in the way of interviews, research, and analysis to reach informed conclusions. Describe the specific methodologies that will be employed to determine the right answers.
SCHEDULE:	Provide a Work Plan with identified tasks and the time period involved, including anticipated start and completion times.
QUALIFICATIONS:	Name the individuals who will carry out the tasks. Give their professional credentials and successes in related activities. Make clear that the client is in expert, experienced, resourceful hands.
COST:	List all fees and related charges.
DELIVERABLES:	Make a commitment concerning the Final Report and any other materials that will result from the service.
CONCLUSION/ RATIONALE:	Discuss specific results and benefits that will occur to reward the client for authorizing you to conduct and complete the assignment.
REQUEST FOR A FAVORABLE DECISION BY THE CLIENT:	Close by asking for the client's official go-ahead. You want an order to initiate the contract and proceed. Ask directly, "When do I start?"

Proposals to Publications

The editorial offices of professional and popular publications often request that potential contributors with ideas for articles or reports send them "queries" to determine whether interest exists or not before submitting finished articles. The process is intended to save time for both writers and editors. Many writers with their articles already planned or written often ignore an editor's invitation to "query" on the theory that the finished piece can speak for itself more effectively than a query proposal.

A skillfully prepared query can sometimes in one or two pages run interference for the follow-up article with great success by arousing interest and stimulating the editorial appetite to see the finished piece. The well-done query proposal functions as a literary hors d'oeuvre or appetizer for the feast to follow. Anyway that's the idea, though it may sometimes be more honored in the theory than the practice, especially when what follows isn't a feast and doesn't live up to the buildup.

Thus, with query proposals, a key point is not to oversell and never to promise what can't be delivered. The usual query result is not a promise to use the article that follows but simply a request to see it, because the query makes it sound suitable and intriguing.

The query proposal to accomplish its purpose must do two things simultaneously: 1) inspire sufficient curiosity to gain the editor's attention and request for the article, 2) brilliantly introduce the article in such a way that exactly the right attitude is fostered and the article when perused does not disappoint.

Writer Jordan Young says about queries, "A good query can sell an editor on you as a writer even if he's not keen on the idea you propose. . .Keep your query brief and to the point. Don't relate your life story, describe your desperate attempts to break into print, or mention your marital woes. State your qualifications, mention appropriate illustrations you may have available, and enclose samples of your work."

Short—to the point—appropriate—interest inducing—these are common characteristics of effective query proposals. You don't want more than a page, at the most two. Don't make the query so long you need an executive query to ask if they want to see your full query. The query should be sent to a publication that you have reason to believe will be interested. It should be sales-oriented to make the proposal compelling and alluring. It should sell the idea and you as the person to present it. Example 16 outlines a query proposal format that serves *if* the idea is right and the proposal is properly targeted:

Example 16:   Query Proposal Format

SUBJECT/BACKGROUND:	Discuss the topic and the background and why it will interest readers of the publication.
ARTICLE DESCRIPTION:	Describe the proposed article with emphasis on interesting and original aspects you intend to feature. Make this a short and enticing overture to your special treatment. Whet the reader's curiosity and stimulate an expectant desire to see what you're talking about. Mention any support materials such as photographs that you intend to include.
ACCOMPLISHMENTS:	Refer specifically to any achievements that qualify you as a contributor in the field concerned. If you have published on the subject, list these credits and enclose representative articles.
COMMITMENTS:	Indicate the proposed length of the article and stress its prompt availability. Request a commitment from the publication to give the idea and your work careful consideration.

The query proposal in effect asks permission from a publication for you to show them what you can do. This means that the query itself should represent your best work and give the recipient a strong and honest impression. Certainly, the query should not be carelessly and haphazardly dashed off. Remembering Neil Simon's *Odd Couple*, in query proposals Oscar Madison slovenliness never suitably announces Felix Ungar tidiness or vice versa. The advance announcement of what's to follow should reflect comparable quality and compatible harmony.

Book Proposals

A proposal for a book is in effect a more ambitious and more elaborate query proposal. The same information is supplied in greater detail. A book proposal also delivers background on the proposed subject, a summary of the contents, and details on the background and achievements of the author with emphasis on the author's qualifications for writing about the subject concerned. The book proposal goes into detail as well about the potential audience for the book, with an indication of who needs or wants it and why.

The book proposal endeavors in the same way as the query proposal to interest a publisher in the subject and simultaneously convince the publisher that you are the right person to produce the book. The book proposal should effectively market the book idea to the publisher. The successful book proposal persuades the publisher that such a book is needed, that an audience is waiting, that the commercial prospects are genuine and promising.

Components of a book proposal include:

1. *Subject:* A stimulating description of the topic and relevant background details are provided together with arguments about why a book on the subject is needed, timely, and marketable.
2. *Book Description:* A synopsis of the proposed book covers the contents in ample depth to convey the scope and quality of the whole. The anticipated length, the planned illustrations, and distinctive or original features are included. The description strives to sell the approach taken as both appealing and particularly effective for the subject.
3. *Audience:* Those who will benefit and how, those who will want the book and why, are identified.
4. *Author Qualifications and Background:* Sell yourself as the person to prepare the book with information about yourself and your relevant experience and accomplishments.
5. *Commitments:* Indicate when the completed work will be made available to the publisher.

The book proposal focuses on making a particular subject attractive and establishing you through your proposal as the person ideally equipped to reach the stated objectives.

Proposals from Special Groups

Federal and state programs exist to support and assist particular groups, and participating in these programs generally commences with successful proposals. Individuals with disabilities, minorities, artists, women, and small businesses are among a large number of such groups.

The initial requirement to benefit from the available programs is full acquaintance with the programs and the opportunities.

Everything previously said about reading instructions carefully and following them to the letter strictly applies to this category of proposals. Typically the competition for available funding is intense, and conscientious care in preparing the necessary proposals is essential.

Learning about opportunities for special groups often starts with the groups themselves. Community and state associations of business owners with disabilities, women business owners, and minority business owners, for example, serve as information resources for activists in these sectors. Getting in touch with the nearest appropriate association is a practical initial step. The location of such groups can be ascertained through the public library, the Chamber of Commerce, and sometimes the telephone book. At the federal level, government agencies maintain offices and programs to assist individuals in these special groups.

At NASA, for example, "Special Assistance Programs" exist "to ensure that all businesses have an equitable opportunity to participate in federal procurement." The programs "include various types of preference programs, such as small business set asides and programs exclusively for minority-owned companies, as well as other forms of assistance generally designed to help companies that otherwise might not be able to compete for a share of government contract awards." NASA has a Minority Business Enterprise Program, a Minority University Program, and "makes special efforts to advise women of business opportunities and preferential contracting programs for which they may be eligible."

The NASA Minority University Program seeks research ties with qualified institutions and emphasizes the "submission of scientifically and technically robust unsolicited proposals by universities in an effort to further the agency's mission." These unsolicited proposals to NASA are evaluated on: 1) intrinsic scientific and/or engineering merit; 2) potential contribution to NASA's mission; 3) availability of funds.

The opportunities awaiting women, minorities, and persons with disabilities in federal and state programs are numerous and varied. The invitation issued by NASA is echoed by most other agencies and departments: "All firms are encouraged to become familiar with and take advantage of any of the special assistance programs for which they qualify."

Federal agencies work closely with the Small Business Administration (SBA) to assist small firms "owned and controlled by socially and economically disadvantaged individuals." Federal legislation enables the SBA to enter into contracts with government agencies for supplies and services, and then subcontract the jobs to SBA-approved small disadvantaged firms. Thus, it pays those who may be qualified and who are interested to contact the nearest SBA office for information and guidance.

Groups such as the hearing-impaired should know about the organizations concerned with their needs. The National Institute on Deafness and Other Communication Disorders in the U.S. Department of Health & Human Services, for example, provides grants for research aimed at expanding knowledge of communication disorders. Other groups and organizations perform similar functions in their areas. They often have both targeted and discretionary funds available and waiting for appropriate candidates to submit suitable proposals.

CHAPTER 14

PERSPECTIVES ON PROPOSAL WRITING

"Blot out, correct, insert, refine,
Enlarge, diminish, interline;
Be mindful when invention fails
To scratch your head, and bite your nails."
Jonathan Swift, 1712

"If you think you can, you can. And if you
think you can't, you're right."
Mary Kay Ash
Mary Kay Cosmetics, 1985

At a meeting on R&D opportunities for small businesses, entrepreneurs, and individual scientists, four men with considerable experience among them on writing and reviewing proposals for federal grants conducted a panel. The purpose of the panel was to look closely at proposal preparation for the Small Business Innovation Research (SBIR) Program from both sides of the challenge, the small business side, the federal agency side. Preparers and reviewers sat down together in peaceful council to share ideas and if possible to avoid fisticuffs.

On the panel were a program manager for the SBIR Program at the National Science Foundation, a researcher from a high technology company successful in winning grants, a state government expert with the job of helping scientists and companies in his state win more grants, and a consultant from a firm whose services to business include guidance in proposal preparation.

The SBIR Program has features that make it different in comparison with other grant programs, but the task of preparing strong proposals is much the same. So are the problems. Thus listening in on what the panelists have to say about this demanding topic could be worthwhile. Let's call the panelists Manager, Researcher, Expert, and Consultant.

149

Manager   A proposal submitted to the National Science Foundation should display the same quality of science, content, organization, and writing required of a technical paper for a refereed journal. When an engineer or scientist submits a paper to a journal, his work competes with submissions by many other professionals with comparable accomplishments. In the case of a distinguished journal, only the best work prevails and gets published. In the SBIR Program, only the best proposals succeed. In a representative year over 1,100 proposals were submitted to the National Science Foundation, but only 102 received Phase I awards. The conclusion is inescapable that if you believe strongly in the value of your research project, your proposal preparation effort should not be casual.

Researcher   One year we submitted 16 proposals and won four Phase I awards. We thought a 25 percent win record was poor. We want to do better. How?

Manager   Winning four out of 16 actually is pretty impressive. For everyone entering the competition, better projects and stronger proposals are how to win. Also be careful. In one competition at the National Science Foundation, 300 proposals were nonresponsive to the requirements of the solicitation. These were returned unreviewed to the senders. They wasted a lot of their time and a little of ours for no reason. The basic requirements apply to all, and they must be met, no exceptions, including the 25-page length restriction. It has to be terribly disappointing to get back a proposal you worked hard on because you took a chance and went over the number of pages allowed. Each proposal we receive must stand up to exacting peer reviews outside the National Science Foundation. Please do your homework to prepare a research proposal. It must have sufficient scientific and technical information to be taken seriously in the peer review process. We don't send proposals to reviewers unless they contain enough science to be reviewed as research.

Consultant   There are different stages of proposal preparation for the SBIR Program the same as any other grant competition. You should be methodical and take a step-by-step approach. Some of the things you have to do include:

1. *Keep up-to-date on SBIR Solicitations.* Read the Small Business Administration presolicitation announcement and obtain copies of the solicitations for the different agencies as they appear.

2. *Concentrate on what you know in your proposal.* Go with what you do best. The SBIR Program is designed to produce seasoned scientific research. It is not a learn-as-you-go program.

3. *Perform an information search.* University libraries, colleagues, government data sources should be consulted. Obtain a bibliography of what has been done. Check out what has been done as if you were selecting a topic for a Ph.D. dissertation. The program requires innovation, not repetition of what has already been accomplished by others.

4. *Prepare your proposal right.* Take the necessary trouble or why bother. If it's due Monday, don't assume you can knock it off over the weekend. You have to write it; subject it to a rigorous technical, style, and content review; then rewrite it.

Researcher   Among our 170 employees are about 45 with Ph.D.s. We perform contract research for government agencies and other clients. The average experience of our scientists is 17 years. We know how to prepare and submit technical papers suitable for refereed journals. We concentrated this ability on the preparation of SBIR proposals.

How did we identify appropriate SBIR Programs for our efforts? We studied the SBIR Solicitations from all participating federal agencies. We identified 50–60 opportunities. A screening committee went through these carefully to reduce the list of possibilities to a manageable number. One aspect of narrowing down was to identify a scientist who was enthusiastic about a project and would serve as its "champion." Every project had to have a champion, someone who would give it the necessary extra time, effort, and enthusiasm to make it work.

The original list of possibilities after winnowing left us with 25 projects, and proposal preparation started. Each proposal had an "Author" and a "Proposal Manager." Throughout the preparation process, proposals were subjected at all stages to continuing peer reviews. These reviews were deliberately made probing and relentless. "What do you mean by this?" and "How can you make this claim?" were questions repeatedly thrown back at those preparing proposals who had to answer and defend them. A proposal wasn't finished until it could go unbruised, unbloodied, and remain intact through the review process.

As an incentive to write better proposals, a review committee chose the best proposals. Their authors and proposal managers received cash awards.

When the proposal is written and ready to go, your job isn't done. You have to see that the prescribed submission procedures are followed to the letter and that it gets to the right destination on time.

Expert   Unfortunately most small businesses don't have 45 Ph.D.s on staff. They lack the resources to mount such an ambitious proposal preparation campaign. They have to do it themselves with whatever help they can find. But that doesn't mean their situation is competitively hopeless. Most grant winners don't have staffs of Ph.D. champions and proposal writing teams. They have to do it themselves, and they do, sometimes with great success. Team proposals that don't win may well get beaten out by single individuals with great ideas who can meet the proposal preparation challenge the same as a team — with work. I agree that extreme care and relentless peer reviews are important whether a large team or a lone eagle tackles the job.

Small businesses should make full use of state and federal services available to them, including computer databases. They should seek out kindred spirits at universities in their area and use university resources in their research and proposal preparation efforts. When you take the trouble to look, there's more help waiting out there than you might expect.

Consultant   Yes, help is available from many sources. Someone at a university can probably be found to give your proposal a technical review or to serve as a consultant on the project. Professional assistance in proposal preparation is also generally not hard to locate. Whatever efforts you make are justified by the opportunity. Such opportunities as those for small businesses in the SBIR Program don't come along often. So make those phone calls. Get help when you need it. Prepare and submit your proposals. Pursue those research ideas you know are valuable.

Manager   Protection of ideas is an area that concerns many who submit ideas and those who don't submit because they don't want to reveal something they want to keep confidential. There is no ironclad guarantee of protection possible, but the National Science Foundation's experience in the SBIR Program has been excellent in this respect. We haven't heard of a reviewer absconding with an idea. When material in the proposal is proprietary in nature, we ask that this information be confined to one page in the proposal. Your proposal is entirely yours until we make an award on it. Federal law stipulates that rights to data and software developed under government support by small businesses and universities will be retained by the developers. Protection of ideas and the treatment of proprietary information are issues covered separately by each participating federal agency in its SBIR Solicitation. You should consult the provisions made for protection by each agency you approach.

Consultant   The policy among the participating agencies in the SBIR Program is to protect confidential materials. But caution is a good idea anyway. When you can explain your project well without disclosing confidential information, that's the safest procedure for all concerned.

Researcher   I'm asked how long it takes to write a proposal. There's no answer except that it takes as long as it takes. Proposals differ so much you can't realistically generalize about variables such as time. A practical average for us is about seven person weeks. To meet the due dates, I'm afraid we have a lot of people working a lot of overtime. Putting background information together for a proposal can be extremely time-consuming. It depends how much you know at the start and how much you have to learn by digging it out of the literature and your own records.

Expert   For the complete process from background research and planning through submission of the proposal, a small business intending to submit a proposal should begin at least six months in advance of the due date, or the maximum time the solicitation issue date allows. This is another reason to grab that solicitation the instant it comes out, so you can get started right then.

Manager   You can probably benefit from studying successful proposals and talking with grant winners. The participating agencies have published abstracts of Phase I grant winners with the names and addresses of winners. From these abstracts you can get an idea about what has been funded, and you can contact winners directly or by mail. In new grant program solicitations, some agencies such as the National Science Foundation publish examples of successful proposals as a guide to others who want to try. The main thing is to follow the instructions which are as precise as we can make them. A common question is how much of a departure is allowed from the requirements of the solicitation. The answer, alas, is none. With over one thousand proposals to evaluate, we can't afford anarchy. Please stick to the solicitation.

Researcher   Be prepared for this to happen. You have a great idea. You work hard. You prepare a fine proposal. And you still don't win, because all the money is gone, or somebody beats you to it, or the timing's wrong, or twenty other reasons. When that happens, and it'll happen, the most important thing is to find out what went wrong.

    Remember that debriefings are offered, and you should take prompt advantage of that opportunity to get the comments of the reviewer and find out exactly why your proposal didn't make it. That information will give you a head start the next time out. You can apply the lessons learned, avoid earlier problems, and do better when you prepare more proposals.

Expert   Whatever the results, you should always prepare more proposals.

Manager   In this and other grant programs, we're counting on that. To succeed, grant programs are completely dependent on the arrival of good proposals in a steady and abundant stream. Contribute to full employment. Mine. Send your proposals. Please.

One Inventor, One Product, One SBIR Win

The above account focuses on a technology-based company with long experience and consistent success in applying considerable Ph.D.-power and teamwork to writing proposals that regularly win SBIR awards and other grants from R&D funding sources. The company and its people are pros at preparing effective technical proposals. The other participants on the panel were also experts, and they spoke with authority well worth heeding on the subject of what to do and what not to do when putting proposals together that deserve and get respectful attention from recipients and reviewers.

The SBIR door is also wide open, if not so easy to enter, for much smaller companies with much humbler technical endowments, and even open for tenacious individuals with product ideas they refuse to let wither on the vine.

The SBIR Program has helped sire countless small business success stories with more joining the list each year. A representative success, and one of the most heartwarming, involves a Midwestern machinist who worked for a large manufacturing company. When his daughter had cerebral palsy and faced a lifetime in a wheelchair, he focused his inventive skills on creating an innovative device that would allow her to move about more freely.

The device impressed many others, and the inventor was urged to make it available for disabled children everywhere. He was eager to do that, so other children could share the improvements his daughter enjoyed. He heard about the SBIR Program and launched a one-man crusade to qualify for an SBIR award. He initiated the process by forming a small business, with himself as President and his wife as Vice President. They were both the management and the employees, the captain and the crew.

This inventor-and-start-up-entrepreneur sought and received technical help from a nearby university and from a technical institute. He made presentations and demonstrations to gain public awareness and support. He obtained agreements for cooperation with and participation in the R&D project from schools in his area. He recruited qualified experts to work with him as scientists and consultants, supplying the technical skills and credentials he lacked. He made one of the scientists the Principal Investigator of the project for an SBIR proposal and assumed the role of Project Director himself. A proposal was written to develop and perfect the device and to make it suitable for FDA approval and worldwide commercialization, because a lot of kids need the liberation it provides.

The proposal was submitted in the SBIR competition, and it won a Phase I award from the Department of Health and Human Services. The award was effectively used by the inventor and the associates he had single-handedly recruited to take the device to the prototype stage with commercialization straight ahead when production and marketing capital could be acquired.

The SBIR win helped the inventor and his small company attract interested investors. The inventor paid tribute to the SBIR Program as a great facilitator and prestige builder in his case. Thanks to the SBIR award earned by his first proposal, the company began moving and gained momentum for commercial growth.

He kept up his drive to learn new skills and to keep turning the results of one successful proposal into a thriving manufacturing business producing the children's support device and selling it wherever the need exists. He sought qualified legal advice to patent his invention in the U.S. and other countries. He worked with a venture capital club and obtained professional counsel from the members of the club to write a business plan that was used to persuade investors. To support his family during these entrepreneurial adventures, he kept his job as a machinist; and his daughter continued to grow and to benefit from her father's remarkable invention.

This is an R&D story with a happy ending. The device received FDA approval based on satisfactory compliance with all requirements, and sales began. What is particularly interesting about the case from the proposal writing perspective is the demonstration that one person, who lacks clear-cut qualifications for subsidized federal R&D but who has sufficient incentive, drive, and determination, can go all the way from project to proposal to prototype to business plan to an operating business with products on the market.

You truly don't have to be a "think tank" stocked with bona fide Ph.D.s. Even just one person with a winning idea can keep pushing until the idea emerges from the treadmill of trial and turmoil as a successful product. A key part of the pushing in this case and in most others, whether from high-tech companies with layer upon layer of skill or a single individual struggling to make the most of a device that can help people, is a good proposal.

SBIR Gives Excellence the Award

In 1986 when the SBIR Program was reauthorized until 1993 by Congressional action and Presidential signature, the original features of the program were retained. The comments by the four experts above and the experience of the lone entrepreneur with a personally recruited support team still apply. Thousands of U.S. small businesses prepare proposals and compete each year for a share of what became nearly half a billion dollars annually by 1987 and thereafter continued at that opulent and enticing level.

At a conference on R&D opportunities, Roland Tibbetts, Manager of the Innovation and Small Business Industrial Program at the National Science Foundation, told representatitves of small businesses, "The SBIR results are astonishing with dividends both for small firms and the government. We note increasing excellence in the proposals that come in. You are learning how to write fine proposals."

Roland Tibbetts and other federal SBIR Program Managers emphasize that while many proposals submitted each year are hastily and poorly done, a large number of the SBIR proposals are excellent. Thus the level of competition is high. To enhance your chances of winning an SBIR award, your proposal should be prepared as conscientiously and carefully as possible. "This is a tough competition but an excellent opportunity for a very bright science firm," commented Roland Tibbetts. To qualify as "very bright" a firm or individual must give the proposal writing task all the energy, effort, and time it deserves. If you don't, your competition somewhere will. So turn it around. Make the competition worry about you by submitting your best work.

CHAPTER 15

SAMPLE PROPOSALS

"What's your next step, my friend? If you know, tell me quickly so that I may adopt your method with all speed."
Albert Einstein, 1934

"Writing is part of science but many scientists receive no formal training in the art of writing. There is a certain irony in our teaching scientists and engineers to use instruments and techniques, many of which they will never use in their working lives, and yet not teaching them to write. This is the one thing that they must do every day."
Robert Barrass
Scientists Must Write

This section contains several sample proposals adapted from actual proposals to show the range of possibilities and practices. In some cases, the proposals have been substantially shortened and much of the documentation omitted. What remains is intended to give you an idea about the directions taken and the enormously varied targets. H. G. Wells once observed that he wrote "to cover a frame of ideas." The proposal writer also attempts to cover and package a frame of ideas and make the result appealing and persuasive to a known and what should be a clearly seen and sharply understood recipient.

A proposal must not be the bland leading the bland, as Murray Schumach defined television, but the wise informing the wise. Edward R. Murrow told the U.S. Senate, "We cannot make good news out of bad practice." Good proposals cannot be made from bad, careless, and uninformed practice. Let these sample proposals inspire you not to go and do likewise, but to go and do better, to prepare real proposals of your own for real targets that win.

Proposal to DOE for a Grant

Following are sample elements (with fictional entries) for a Department of Energy "Appropriate Technology Small Grants Program" proposal. The closing date for receipt of proposals was March 26, 1981; and the results of the program, which is now history, have been published. The program distributed over $25 million in close to 2,200 small grants. The goal was to further the use and commercialization of appropriate technologies—technologies that use renewable resources for energy such as sun, wind, and water, with emphasis on self-help efforts and development of local resources for energy applications.

The solicitation for the program was well designed to be used by many people who had never submitted proposals before—homeowners with do-it-yourself projects, farmers developing biomass energy systems, nonprofessionals with practical energy saving ideas. The solicitation contained easy-to-interpret proposal preparation and submittal instructions, and included proposal forms to be filled in by applicants. The first page (partially reproduced below) facilitated computer processing and routing. The rest of the proposal was typed on tear-off pages from the solicitation.

The sample elements illustrate one proposal format used by a federal department in a national grant program. The ready-to-use forms made the proposal simpler to prepare than in many other programs.

The project and other information on the sample elements are invented, although grants were made in the program for various solar devices of a similar nature. The fictional entries are not as detailed or complete as they would be in an actual proposal, but rather are designed to show what a proposal format might require and in brief how a response might be commenced. Among the approximately 2,200 winners in this competition were many who had no experience writing proposals but who had sufficient confidence in their ideas for appropriate technology developments or demonstrations to try.

Appropriate Technology Small Grant Application Proposal Forms (Fictional Entries)

Project Title: _Design and Develop a Solar-Heated Dryer for Firewood_

If this proposal is submitted by an individual:

Applicant Name (print or type) _Jerry Maloy_

Signature _Jerry Maloy_

Date: _March 9, 1981_

DATA FORM — 1981 (Use only these blocks — abbreviate if necessary.)

(Circle One): (MR) / MS / MRS / MISS / PROF / DR

SECTION A

LAST NAME OF APPLICANT

| M | A | L | O | Y | | | | | | | | | | | | | |

FIRST NAME OF APPLICANT

| J | E | R | R | Y | | | | | | | |

HOME AREA CODE & TELEPHONE NUMBER

| 2 | 2 | 2 | 3 | 3 | 3 | 4 | 4 | 4 |

MAILING ADDRESS

| G | R | E | E | N | | P | A | S | T | U | R | E | S | | F | A | R | M |

CITY

| R | U | R | A | L | | S | U | N | S | H | I | N | E |

STATE | T | X |

ZIP CODE | 7 | 9 | 0 | 7 | 2 |

BUSINESS AREA CODE & TELEPHONE NUMBER

Jerry Maloy
APPLICANT'S SIGNATURE

MARCH 9, 1981
DATE

SECTION B

GRANT CATEGORY

B - | B |

TOTAL GRANT REQUEST (to nearest dollar)
(From Page A-10a, Line 13)

| 1 | 2 | 5 | 0 | 0 |.| 0 | 0 |

SECTION C

Choose the ONE applicant identification category that BEST describes you and put the corresponding number in the box.

C – 1. Individual(s)*
C – 2. Small Business
C – 3. Indian Tribe
C – 4. Non-profit organization or institution
C – 5. College or university
C – 6. State or interstate agency
C – 7. Local or regional agency

C – | 1 |

*NOTE: If 1. is checked, and there is no co-applicant, go directly to next page.

A. Project Summary

This summary may be used to assign your proposal to the proper reviewer.

1. *From the list provided, indicate the goals that will be the focus of the project*:

    (a) The use of renewable resources and the conservation of nonrenewable resources.
    (b) Applications which are energy conserving, environmentally sound, small scale, and low cost.
    (c) Applications which demonstrate simplicity of installation, operation, and maintenance.

2. *In 150 words or less, provide a nontechnical description of what you plan to do and the expected results*:

A house furnace, installed at this farm in 1979, uses wood fuel taken from forestland on the property. Introducing wood fuel for home heating has reduced consumption of natural gas in the house about 50 percent.

A major problem with firewood is that green wood takes over a year in the open air to become dry enough for efficient use as fuel. The same problem is common all over the United States.

The goal of this project is to design and develop a practical firewood dryer that uses solar energy for drying heat. This solar-heated dryer will make it possible to dry wood sufficiently for utilization as fuel in a few weeks, saving months over conventional air drying.

The device will be one homeowners and farmers can build themselves following the scale drawings and step-by-step instructions that will be developed in the project.

B. Project Results

1. *Using a few key words and phrases (instead of complete sentences), describe what devices, systems, or information will exist as a result of your project.*

    (a) Solar-heated dryer for firewood operating at Green Pastures Farm.
    (b) Scale drawings of the dryer.

(c) Detailed list of required building materials and construction tools.

(d) Step-by-step instructions for building the dryer.

2. *How will others use what you have developed or learned?*

The solar-heated dryer will be designed for materials, tools, and construction methods others can easily use to make firewood dryers capable of meeting home or business fuel needs.

Persons with construction skills or novice builders with professional help will find this a practical way to solve the problem of making firewood dry enough to burn without long delays.

The U.S. has an estimated 14 million wood-burning facilities consuming about 45 million cords of wood annually. Most of these wood-burning facilities are homes, farms, or individual establishments. The solar-heated dryer for firewood could significantly benefit a large number of people now using wood as a heating fuel.

C. *Technical Description of Project*

New ideas on the design and construction of a solar-heated dryer suitable for drying firewood will be incorporated in a full-scale, functioning prototype. Scale drawings and instructions will be completed that others can follow to build identical models.

The original features of this dryer will be based on the applicant's experience building solar greenhouses, sunspaces, and drying devices as well as his experiments with different materials and designs.

The work plan will be carried out in five steps over a period of 11 months. The five steps of the work plan are:

1. Preliminary Design
2. Experiments with Materials and Shape Configurations
3. Experimental Model Construction
4. Final Design
5. Prototype Construction

(*Note*: The technical description for this proposal would then continue for five pages or less with specific details on each of the steps in the work plan. The description would emphasize features that are given in the solicitation as criteria to be used in judging by technical reviewers.)

D. Social and Economic Benefits

This project to develop a solar-heated dryer for firewood will be carried out entirely by the applicant using local materials and resources. This is done to assure that the resulting solar device is one that individuals with initiative and self-reliance can construct for themselves anywhere.

Wood was the common fuel in America until cheap fossil fuels became dominant in the 20th century. Today with fossil fuels no longer cheap and the supplies dwindling, wood is again popular for home heating and related applications. Wood suitable for firewood is found all across the United States, and millions have already turned to nature's number one renewable energy resource by using wood as fuel. Millions more are certain to follow.

Wood must be dry to burn well, produce maximum heat, and minimize such problems as air pollution and creosote buildup. Green wood takes a long time to dry, over a year in many parts of the U.S. The solar-heated dryer that will be developed in this project will reduce wood drying time to a few weeks. The dryer will greatly benefit those already using wood fuel, and it will make the use of wood fuel much more convenient for others.

This solar device will facilitate the switch to wood fuel, thus stimulating development of local energy resources, reducing dependency on imported fossil fuels, and helping individuals achieve a greater degree of energy independence. The device will also readily lend itself to local commercial activities, with builders constructing the dryers for local residents and businesses.

(*Note*: In an actual proposal, this description would continue with point by point attention to the specific criteria listed for consideration by reviewers.)

E. Qualifications of Key People

The applicant will work alone in the project to design and build a prototype solar-heated dryer for firewood. He expects to devote 600 hours to the five tasks of the project.

The applicant holds B.S. and M.S. degrees in agriculture and agricultural engineering from the Southwestern Institute of Technology. While attending college, he worked part-time as a draftsman and took graduate courses in engineering design. He has the professional skills needed to prepare the scale drawings for this project. He is also experienced in construction work, including a barn, tool shed, home additions, and other structures at Green Pastures Farm.

(*Note*: Some proposal formats would require the applicant's resume and the resumes of other project personnel at this point in the proposal or in an appendix.)

F. Project Schedule

List the major activities you plan during the performance of your project. For each activity, show the start date, the completion date and any important intermediate milestones.

EXAMPLE

S = Start
C = Complete

Steps	Months after Grant Award												
	0	1	2	3	4	5	6	7	8	9	10	11	12
Purchase Materials		S—	—	—C									
Build Model									S—	—	—	—	—C

Steps	Months after Grant Award												
	0	1	2	3	4	5	6	7	8	9	10	11	12
Preliminary Design		S—	—	—C									
Experiments with Materials and Shape Configurations		S—	—	—	—	—C							
Experimental Model Construction				S—	—	—	—C						
Final Design						S—	—	—	—C				
Prototype Construction									S—	—	—	—	—C
SUBMIT FINAL REPORT										S—	—	—	—C

PERIOD OF PERFORMANCE

Grant projects are expected to be completed within 12 months after award. If you propose a longer period of performance, furnish a schedule and statement, justifying the extended period in excess of 12 months.

G. *Budget*

BUDGET SHEET

TOTAL FUNDING REQUESTED

$ 12,500.00

Cost Category Number	Cost Category	Estimated Weekly Hours	Hourly Rate	Estimated Number of Weeks	TOTAL
1.	Salaries and Wages (list name or position)				
	(1) Jerry Maloy	40 x	$25. x	15	$15,000.00
	(2) _____	___ x	___ x	___	_____
	(3) _____	___ x	___ x	___	_____
	(4) _____	___ x	___ x	___	_____
	(5) _____	___ x	___ x	___	_____
	(6) Others (list names and other data on Budget Support Sheet)	___ x	___ x	___	_____
			Total Salaries and Wages		$15,000.00
2.*	Fringe Benefits (if applicable)				_____
3.*	Equipment				_____
4.*	Materials				3,250.00
5.*	Supplies				100.00
6.*	Travel				_____
7.*	Subcontractors/Consultants				_____
8.*	Other Direct Costs				150.00
9.	Total Direct Costs				$18,500.00
10.*	Indirect Costs				_____
11.	Total Project Costs				$18,500.00
12.*	Proposed Cost Sharing				6,000.00
13.	Total Amount Requested from DOE (11 minus 12)				$12,500.00

ADVANCE FUNDS REQUIRED:

Indicate the amount of advance funds necessary for the first 4 months of the project (refer to Section IX of the Program Solicitation). $5,000.00

*Itemize on Budget Support Sheet, pages A-10b, A-10c, A-10d, and A-10e.

(*Note*: In the full proposal, the foregoing Budget Sheet would be followed by Budget Support Sheets giving cost details on materials needed for the project. For this hypothetical project, the Budget Support Sheets would also show that the applicant intends to contribute 40 percent of his personal labor as Proposed Cost Sharing since the evidence of cost sharing significantly strengthens a proposal. A number of Budget Sheet items here are blank because the applicant will do the work alone, no equipment purchases are necessary, no travel is required, and indirect costs are not allowed because the applicant is an individual.)

H. Patent Waiver

If you want the Government to give title to inventions back to you, please check the box below.

☑ Small Business or Individual—I qualify as a small business or an individual and request that title to inventions be given back to me.

(*Note*: Applicants that do not qualify as a small business or individual who want title to inventions returned to them are required to explain their plans for commercialization of the invention by manufacturing and marketing or by licensing others to use the invention.)

This Appropriate Technology Small Grants Program solicitation ends with a very useful proposal checklist to guide applicants in making certain everything has been done. The checklist refers to different proposal elements and asks specific questions about contents so the applicant can decide whether or not he has prepared the proposal accurately and included the information that reviewers will look for during evaluation.

Such a checklist is an excellent means of self-protection against mistakes or omissions in the proposal and a good way to determine if you have been responsive to the solicitation. Go through the checklist carefully, and you can decide if you paid enough attention to the criteria used for judging proposals. When you find something that needs fixing, make the changes immediately.

Many solicitations and RFPs do not provide such checklists. When responding to those programs, you should set up a suitably detailed checklist of your own with which to check your proposal when you think it is ready to go. Such steps pay off in greater accuracy.

Sample Proposal: A Magazine Series on People with
Disabilities and Mainstreaming

Unsolicited proposal from an educator-researcher to a parents magazine
concerning the need for a series of articles. Adapted and condensed from
an original proposal by Deafness Consultant, Helen E. Meador, Ph.D.

Meeting Individual Needs and Mainstream Challenges

[BACKGROUND] I am an educator-researcher with a Ph.D. in the field
of Deafness Communications and have extensive experience. I serve as a
Commissioner on my community's Commission on Disability Issues.
You published recent articles highlighting issues evolving from Public
Law 94–142, which requires the least restrictive educational environment
for all children, and from the Americans with Disabilities Act of 1990,
which requires reasonable accommodations for handicapped individuals.
Thank you for these insightful and useful articles which demonstrate
awareness, sensitivity, and willingness to deal with an extremely complex
problem, that of differing abilities among children as well as the attitudes
and responses of parents. Your articles are a good start. Now vigorous
follow-up is needed.

Love Overcomes Communication Obstacles

[PROBLEMS IDENTIFIED] Your articles on differing abilities promote
greater understanding and deserve acclaim. However, I have observed
serious gaps that need attention. After identifying these gaps, I want to
propose a solution for your consideration.

In a 1991 article on family relationships, an allusion to Deafness was
regrettably misleading. Joy was expressed that the author's children were
not born Deaf because an essential bond between parent and child is the
"umbilical cord of the voice." Hearing parents of Deaf children might
read this cruel statement and weep. They should be told that it is not
accurate. Love, not the human voice, is the essential bond between parent and child. The bond of love can be communicated in many ways that
need not involve hearing and the voice.

One case among many illustrating this truth was that of my pro-
foundly Deaf sign language instructor at New York University. He was
born Deaf to hearing parents. His mother developed and used home-
made signs to communicate with him, and his father's love was success-

fully communicated without signs or the human voice. He heard no sounds, but family communications were excellent, because he was loved.

Deafness in a hearing family (90 percent of hearing-impaired children are born to hearing parents) need not cause insurmountable difficulties. Understanding, accurate information, and avoidance of misconceptions such as the one concerning Deafness in your article are the answers. This publication is in a strong position to provide current, accurate information on Deafness in families and to serve the vital cause of handicapper understanding both specifically and broadly.

Monthly Articles Proposed To Address the Problem

[SOLUTION] I recommend that an urgently needed regular monthly series of articles be published by your magazine portraying individuals with differing abilities in a variety of family and educational settings. This would be in harmony with your noteworthy attempts to keep abreast of developments in all areas of social and educational importance.

Articles on Deafness and other disabilities that are too often misunderstood and too seldom understood will be covered in the series. Thus, your magazine and its readers will remain current and well-informed about our increasingly visible and voluble differently-abled populations as they step up their activities in mainstream society. Monthly features on these populations will attract new readers among parents whose needs and concerns are not being served at this time by popular periodicals. Mainstream children to succeed must have mainstream parents who are conscientiously informed and supported by community understanding.

I predict that such a series will attract a wide and appreciative readership and significantly add to your prestige as a bold and pioneering publication among popular periodicals.

In just the Deafness area, the opportunities and challenges are immense. At America's schools, over 30,000 hearing-impaired children are mainstreamed. About 55,000 hearing-impaired children are part of U.S. educational settings that may or may not be mainstreamed. These children and their families represent a large, significant, and too often neglected group.

[PROPOSED DEAFNESS ARTICLES TO START THE SERIES] (The proposal supplies details on eight areas of informational need for parents of hearing-impaired children and suggests that these offer an excellent starting point for a hard-hitting, high-impact series.)

The Availability of Proven Expertise for the Series

[PERSONAL QUALIFICATIONS] My qualifications, experience, and commitment equip me to contribute articles in the areas of Deafness and other handicapper needs, issues, challenges, and opportunities. I am also able to keep you abreast of the field and to apply inside knowledge when selecting appropriate topics for the series.

My skills and experience in Deaf Education and Communication are documented by the accompanying Curriculum Vitae. As a classroom Teacher of the Deaf, I became familiar with the communications problems of Deaf children, their families, and their hearing peers. As a tutor of Deaf adults and participant in Associations for the Deaf, I learned firsthand about the daily struggles of the Deaf in mainstream America. As Assistant Director on a Consortium for Enabling Technology, I established dialogue among handicapper groups and worked closely with them in solving problems. As a Deafness Consultant, I have access to university facilities and close ties within the field of Deaf Education and the Deaf Community. As a member of the Commission on Disability Issues, I have direct knowledge about needs within both schools and communities.

The Ability and the Resources

[REQUEST FOR AN OPPORTUNITY] I am well-qualified to deliver a regular feature article for this magazine starting with the proposed Deafness series and following with monthly articles on disability needs and issues. The articles particularly will focus on children with differing abilities and cover authoritatively and interestingly what happens, and what should happen, when they are mainstreamed. The subjects of the articles, their technical accuracy, the continuing debates about mainstreaming, as well as unsettled and unsettling handicapper questions will attract large numbers of readers and inspire frequent and earnest reader response.

My record as a writer and academician shows that I am scrupulous about meeting deadlines. I have an effective network of contacts available for expert guidance and information when needed. I am prepared to commence the series immediately upon receiving an indication of your receptive interest.

The Proposal Package Submitted to the Magazine

Appropriate support and backup materials with the above proposal include personal documentation, professional history and bibliography, writing examples, and detailed discussion of potential articles in the proposed series.

The objective of the proposal is to highlight a challenging area of need — that is also an area of opportunity — for the magazine, and to initiate the process of obtaining a hearing for the person submitting the proposal as a contributor of periodic articles on Deafness and the handicapper communities. Note that the proposal includes emphasis on the increased readership and prestige such a handicapper series will bring the publication.

A proposal of this type should be followed up by further contacts, submission of sample articles, supplementary arguments, and complementary letters as necessary to attain the projected goal. An unsolicited proposal is an initial marketing vehicle that launches but does not necessarily complete a sales campaign. The campaign ends when the sale is made.

Sample Consultant Proposal: Deafness Awareness and
Communications Seminar

Solicited proposal from a Deafness Consultant to an organization.
Adapted and condensed from an original proposal by Deafness Consultant, Helen E. Meador, Ph.D.

Proposal for an On-Site Seminar:
Deafness Awareness and Communications

Meeting the Special Communications Needs of
Hearing and Hearing-Impaired Staff

Thank you for the recent opportunity to discuss the problems you are
encountering in achieving effective communication among Hearing and
Hearing-Impaired staff members. The difficulties and stumbling blocks
you report are common ones that can be overcome by assisting Hearing
personnel in acquiring deeper knowledge and understanding of Deaf
Culture, and guiding Deaf personnel in comprehending Hearing Culture.

Learning sign language, the language of the Deaf, will alleviate some
problems; but signs alone are not enough. Appreciation of Deaf Culture
and its profound differences from Hearing Culture, including politeness
issues and eye gaze, is equally critical. If Hearing and Deaf staff members
share sign language and resolve politeness issues from their mutual
perspectives, the communications barriers now inflicting trouble and
confusion will crumble.

Deafness Awareness and Communications Seminar:
Specially Designed to Eradicate Impediments and
Help Achieve Harmonious Employee Relations

Mastery of basic sign language and increased sensitivity to politeness
and cultural issues for all concerned will reduce tension and improve
working conditions by facilitating better communications and producing
a friendlier work environment. The end result will be enhanced productivity for your company and a much happier staff with better morale.

As the best means of achieving these results, I propose a Deafness
Awareness and Communications Seminar specifically created for your
organization and its special needs.

Schedule: The seminar will consist of weekly two-hour sessions for a period of six weeks. The six seminars will be conducted at your location and the times will be selected to accommodate staff schedules and availability. **Participants:** My advice is to encourage all employees, both Deaf and Hearing, to attend the sessions. The sessions themselves will be designed to begin the process of removing obstacles and misunderstandings between the two groups. **Consultant Fees:** (A fee schedule based on the number of participants and stipulating the Consultant's minimum was included. The statement was made that seminar handouts were covered by the fees.)

Seminar Features

- The Hearing staff will learn the basics of communicating effectively with Deaf colleagues.
- Sign language instructions will focus on the sign language actually used in the Deaf Community.
- Participants will learn the American Manual Alphabet for handspelling to allow expanded communications with Deaf colleagues.
- Coworkers will master basic sign language greetings to achieve warmer communications and create a more cordial and companionable work atmosphere.
- Hearing employees will be informed about and made comfortable with special politeness factors among Deaf people that differ from mainstream America. Deaf employees will also receive instruction concerning this difference with practical suggestions for understanding and getting along with their Hearing coworkers.
- Representative samples of writing by Deaf people will be examined and instructions given in interpreting the variations that sometimes occur from standard English.
- Participants will receive useful handouts on Deafness issues, reading lists, other aids, and helpful advice.
- Staff members will gain greater awareness and understanding of Deafness, and overcoming old barriers will bring personal and company benefits from the seminar sessions.

Among Deaf Culture issues covered will be attention to such key questions as these: How do Deaf people let others know they want to communicate? How do they use intense eye gaze, thus violating the Hearing taboo against "staring?" How do they vent anger, tell jokes, use the telephone, "call" across the room, "whisper," "yell?" How do they feel about "secrets?" And so on.

Each presentation will occur informally with questions encouraged. Time will be allotted at each session to review the information covered, to discuss existing misconceptions, and to work actively at removing barriers. The third session will end with a quiz and an opportunity to write comments about problems confidentially. These confidential comments will be addressed in the last three sessions. The sixth session will end with a cumulative quiz and completion of seminar evaluation forms that are delivered to the company.

[Note: Problems and issues discussed during the six sessions of this seminar will be held in strictest confidence according to the State Code of Ethics (copy attached).]

Consultant Experience and Qualifications

I am a Deafness Consultant specializing in Deafness Communication and have wide experience as a teacher and researcher. My professional accomplishments and academic attainments are documented by the accompanying Curriculum Vitae. I hold Degrees from the City University of New York, New York University, and a Ph.D. from the University of Michigan. I am a Certified Teacher of the Deaf, a lecturer on Deafness Issues and Culture, and a Sign Language Interpreter. I have extensive experience working with the Deaf Community on many levels. I worked in the theatre as a performer and consultant and facilitated communication among Deaf and Hearing actors. I discussed Deafness and its ramifications on radio and television.

Special Role of Management and Staff Participants

The success of the seminar will depend on active support from and participation by company executives. Staff participants must be fully aware of this support, preferably shown by the involved presence of managers and executives at the sessions. Hearing participants will be strongly encouraged to share their fears, problems, and insights about communicating with Deaf people, and Deaf employees will receive similar encouragement concerning communication with their Hearing co-workers.

As the Seminar Facilitator, I shall make continuing efforts to identify specific communications needs at your company and to deal with them directly through action and discussion at the sessions. I emphasize that it is *your* seminar and particular attention should be given to the needs, problems, and concerns now present.

Progress Assessment Meetings with Management

I propose to meet separately with managers and executives after the third session to discuss progress and to pinpoint issues that can be addressed in the second half of the seminar. A follow-up meeting with managers and executives after the sixth session will include a summation of results, identification of remaining trouble areas, and counsel for the future.

When the First Session Will Commence and the
Benefits You Might Expect

The seminar will begin at an agreed upon time within two weeks after your acceptance of this proposal. Please contact me if you have any questions concerning the proposal. I am prepared to make changes and adjustments in accordance with your needs and schedule.

You can expect much-improved relations between Deaf and Hearing staff people to result from this seminar. The catalytic effectiveness of greater knowledge and understanding in bringing the two groups harmoniously together is widely confirmed. Your workplace environment and staff morale will definitely improve. The firm will substantially profit from the instruction and from the cooperative reaching-out among employees that is steadily emphasized and accomplished during the six weekly sessions.

I am ready to proceed. Please let me know when you wish the seminar to start.

The Consultant Proposal Package

Together with the above seminar proposal, the consultant submitted to the potential client a descriptive brochure on the Deafness Awareness and Communications Seminar as well as personal documentation confirming the statements about experience and qualifications made in the proposal.

Note that while the subject matter of this proposal and the previous proposal for a series of magazine articles concerns aspects of Deafness, the contents and points emphasized differ greatly because the purposes of the proposals and their targets are quite different.

Who gets a proposal always determines *what* a proposal includes and *how* it is prepared. A proposal is carefully planned and written to reach a specific target, and the identification of the target must be known before the proposal is prepared.

Sample Proposal for a   Old-Time Radio Programming at
Promotional Idea: Contemporary Radio Stations

Unsolicited Proposal to Radio Station Managers and Program Directors [Adapted and condensed with permission from an original proposal by Lawrence Rao, collector and programmer of vintage radio performances.]

Invitation to Expand Your Audience and Your Audience's Pleasure With Old-Time Radio Programming

Vintage Radio Programs Win New Audiences of All Ages

Why include vintage radio broadcasts in your schedule? Answer: Because vintage radio consistently proves a surefire method of increasing your ratings and ad sales. Listeners *Love* vintage radio series as repeatedly confirmed by audience surveys. You will love them too — and their ability to deliver.

- A 200+ percent increase in a station's evening audience.*
- A 600+ percent increase in 19–34 age group listeners.*
- A 1100+ percent increase in the teen audience.*

Introduce *all* audiences to the hilarious comedy of Burns & Allen, Jack Benny, and Fred Allen; the gripping adventures of "Escape;" the never outdated mysteries of "Sam Spade," "Nero Wolfe," and "Nick Carter;" the out-of-this-world thrills in "Dimension-X;" the goose bump suspense of "Inner Sanctum" and "Suspense." Sounds great, is great!

*Such dramatic audience increases and others coast to coast were documented by radio stations in Los Angeles, Chicago, and Boston [see the accompanying report of audience figures]. During a decade of declining AM ratings and AM ad sales, classic radio broadcasts from the public domain often revitalize AM results in ratings and sales. Your call letters will become the entertainment bright spot on the dial when you introduce these Golden Memories from the Golden Age of Radio.

*Old-Time Radio—Proven Method for Improving Your Station's
Market Rating*

As a radio professional, you work to keep programming not only
competitive but far ahead of other stations. This unending challenge calls
for progressive vision and new ideas. One of the best new-time ideas
today is innovative programming based on Old-Time Radio.

Classic broadcasts from radio's Golden Age, those never forgotten
shows from the 1932–1962 era boost market ratings, delight and expand
the audience, please and attract advertisers. As ecstatic listeners phone
and write your station to express gratitude and request favorite vintage
radio series, advertisers will line up with ad dollars to sponsor not just
single nights but extended series. It's happening right now in many parts
of the country. It will happen in your area.

All-Time Stars in All-Time Favorite Shows

The list of famous voices and celebrated series available in my collection for broadcasting on your station is long and irresistible.

Stars, Stars, Stars

Orson Welles, Laurence Olivier, Olivia De Havilland, Errol Flynn,
James Cagney, Rita Hayworth, Alan Ladd, Burns & Allen, Fanny Brice,
Eddie Cantor, Roy Rogers, Jack Benny, Ray Milland, Bette Davis, Kirk
Douglas, Jane Wyman, Ronald Reagan, Bing Crosby, Gene Kelly, Rosa-
lind Russell, Gregory Peck, Vincent Price, Dick Powell, Mike Wallace,
Van Johnson, Charles Boyer, Clifton Webb, Gene Tierney, Lana Turner,
Fred Allen, Judy Garland, Tex Ritter, Joan Blondell, and many many
more.

Immortal Series

LIGHTS OUT / MERCURY THEATER OF THE AIR / JACK
BENNY PROGRAM / FRED ALLEN SHOW / ESCAPE / BOB
HOPE SHOW / PHIL HARRIS & ALICE FAYE SHOW / THE
WITCH'S TALE / 2000 PLUS / MURDER AT MIDNIGHT/ GULF
SCREEN THEATER / LUX RADIO THEATER / STRANGE DR.
WEIRD / CAVALCADE OF AMERICA / BEYOND THIS WORLD /
BEYOND TOMORROW / DIMENSION-X / SUSPENSE / MY
FRIEND IRMA / SCREEN DIRECTOR'S PLAYHOUSE / DICK
TRACY / INNER SANCTUM / MURDER BY EXPERTS / NERO
WOLFE / SAM SPADE / PHILIP MARLOWE / RICHARD DIA-
MOND / and more more more!

All the vintage radio episodes offered are in the public domain. Each is carefully selected with current ethnic, religious, and lifestyle mores considered. Programs that are now considered offensive by some American listeners are not included. You need not worry that these vintage radio selections will ever offend any ethnic or religious groups in your audience.

Recommended Scheduling and Program Lease Arrangements

(The proposal included details on daily and weekly programming schedules for the vintage radio materials and leasing fees for the cassettes involved that will be supplied to the station and returned to the owner.)

The proposal discussed the high quality of the cassettes and of the sound. The radio station was informed that the vintage broadcasts could be hosted by station announcers or by the supplier who emphasized his extensive experience in programming and presenting such broadcasts. The special qualifications of the collector and vintage radio expert as a radio historian and media consultant were described.

The programmer recommended a Monday-Friday scheduling arrangement with a different radio theme each night, including drama, comedy, science fiction, adventure, mystery. The production of special Old-Time Radio Festivals was proposed in cooperation with the station and reflecting the station's particular needs.

The Old-Time Radio Proposal Package

The proposal was carefully packaged for potential radio station clients as an *Old-Time Radio Book*. In addition to the written proposal, an impressive array of support materials was added including radio station documentation figures on the effectiveness of vintage radio as a means of increasing audiences, a partial listing of the vintage shows available for programming, and a 90-minute sampler cassette tape featuring a large number of brief segments from the recommended shows. The station was invited "to broadcast portions of this cassette and encourage listeners to phone or write the station about their interest in hearing Old-Time Radio." Background articles included newspaper reports on the programming: "To some listeners, it's nostalgia. To others, it's one of the purest art forms, the theater of the mind." ". . .wide appeal to young and old audiences. After a series has completed its run, it is put on the shelf and recycled later in the year. The ratings prove the formula works."

The proposal concluded with instructions about how the radio station should take advantage of the opportunity to "allow public domain vintage radio broadcasts to enhance market ratings, increase advertising, and delight audiences." Ways to make contact by telephone or mail for follow-up meetings and follow-on discussion were stated clearly and repeated several times in the package.

Remember, if your proposal idea is compelling but they can't reach you to explore it further, you lose the opportunity.

This was a complex and elaborate proposal package to describe and sell an idea and to recruit radio stations as regular clients for the collector's vast number of shows and programming options. The cassette tape of program excerpts was a particularly strong and effective complement to the written proposal.

The proposal demonstrates what frequently must be done to convince proposal recipients that an idea and offer have merit and deserve attention and prompt action.

Sample Unsolicited Internal Company Proposal:
[Fictional entries are used for illustration.]

MODERN WARES & WAYS, INC.
El Dorado, New Mexico

May 25, 1991

To:	Carl La Fong, Executive Vice President
From:	Enoch Zagetta, Personnel Manager
Subject:	Proposal to combat costly system fear by improving employee familiarity with and acceptance of our new computer system and software in the interest of greater effectiveness and efficiency.

Purpose and Problem

When MWW acquired its $2.2 million computer system with a central computer and multiple terminals on November 19, 1990, the objective was to have the system functioning smoothly in all departments within six months. Our plan was improved productivity and greater efficiency throughout MWW with accounting, personnel, parts and inventory control, design, product development, and production operations on-line and the MWW system contributing to accelerated company growth.

At the six-month point, progress has been made in some areas, but not nearly enough. Many employees are still fearful of the system and reluctant to learn how the instrumentation can make their jobs and lives easier. Some employees are fearful that if they let the computer help them, the computer will soon take over their jobs. We have a real personnel and morale crisis here. About half of the workers avoid using the system whenever possible and stick to old methods like flies to flypaper.

Collectively they seem to view our system the way a character sees computers in an Eric Bogosian skit: "All these computers are talking to each other, man. . .What are they talking about? I'll tell you what they're talking about. They're talking about you and me. . .how to use *us* more efficiently."

Our MWW people are afraid of being used. They worry about their capabilities with the new technology. The brief and cursory training sessions given by the supplier at the start simply didn't do the job. We have to take positive action in this situation, or this super-system will continue to be a multimillion dollar albatross.

Proposed Solution to the Problem

I propose immediately launching throughout MWW a comprehensive new computer system indoctrination and training program called *Let Your Computer Be Your Pal*! The goal is to help our people overcome fear, get on a comfortable basis with these devices, and learn that computers are friends, not foes. [Mr. La Fong, are *you* getting maximum benefit from the terminal in *your* office? Yes, sir, the program I propose will include you and other MWW executives. Companywide means companywide, and this program soon will make us all computerwise.]

Program Details

After consultation with MWW department heads, employee groups, and several individual workers, I recommend the following:

• Eight "Computer Get Acquainted" Sessions for each terminal operator. The sessions can be arranged at El Dorado Community College (EDCC) in the Computer Science & Information Center. EDCC has agreed to organize the sessions to fit our system and needs with emphasis on start-up fundamentals in step-by-step fashion. **Cost**: EDCC will charge the company $450 per employee for the eight-sessions. Estimated number of participants: 75.

Total Cost: $33,750

• Computer experts Wanda Miller and Judith O'Leary of EDCC have agreed to work on a consulting basis at MWW directly with employees in each MWW department to answer questions, solve problems, demonstrate options, and guide our people in "becoming pals with the system." **Cost**: Consulting fees for Miller and O'Leary are $35 per hour. Estimated Need: 100 hours.

Total Cost: $3,500

• "After Hours Happy Hacker Happenings." These will be company-encouraged activities in which terminal operators after work and on weekends continue "hacking" on their own, learning new tricks of the trade, and getting on better terms with the equipment. The company will provide refreshments from the MWW Cafeteria for these volunteer occasions that will otherwise be uncompensated. **Cost**: Estimated Food Expenses, 100 meals weekly for six weeks, 600 meals at $4.75/meal.

Total Cost: $2,850

[*Note*: My belief from talking with department heads and employees is that most operators will volunteer to upgrade their skills and give some of their time to help the company benefit from this expensive system after hours and in the EDCC course. If the volunteer approach doesn't work as expected, I recommend paying employees straight time to participate.]

• "Incentive Reward Program." Employees who complete the EDCC course, cooperate fully, practice on their own, and develop the requisite computer skills will receive Two Extra Vacation Days or equivalent pay at the end of the program. **Cost**: 75 employees @ $400/employee.

Total Cost: $30,000

Program Cost Summary

External Costs

El Dorado Community College Courses Eight Weekly Sessions 75 Persons @ $450/Person	$33,750
Consultants Available On-Site at MWW 100 Hours @ $35/Hour	3,500

Internal Costs

Cafeteria Expenses, 600 Meals @ $4.75/Meal	$ 2,850
Incentive Reward Program 75 Employees @ $400/Employee	30,000
Total Estimated Cost	$70,100

I submit that this investment will enable us to start the process of gaining full benefit from the computer system. This program alone will not make everyone comfortable with computers and efficient in their use, but it will take a substantial step toward that objective and will effectively demonstrate that the company is very serious about the computerization effort.

Conclusion and Implementation

We have a computer system at MWW with great potential that is not currently being achieved. The proposed program at a cost considerably less than the heavy cost of expensive, underutilized, sometimes idle equipment will initiate a continuing process of making MWW employees computer enthusiasts.

The sessions at El Dorado Community College can begin with three weeks notice. Instructors Miller and O'Leary are available to commence immediate consultations with our personnel at MWW during times that do not conflict with their EDCC schedules.

Plans for the After Hours sessions are ready. I can announce the details to employees and commence the program upon approval. Please let me know if you have any questions or suggestions about these urgently needed steps. I am available to discuss the plans and their implementation whenever it is convenient for you.

Sample Book Proposal:

Book Proposal

Technological Innovation and Small Business in America

by A.S. Piranti
Lecturer on the History of Science and Technology
[Note: Title and Author are fictional for illustration.]

In the global marketplace of the 1990s and the 21st century, world economic leaders will be those who dominate in achieving commercially successful technological innovations. The triumphs of such innovations in the final quarter of the 20th century again proved this axiom that was repeatedly demonstrated in earlier 19th and 20th century developments.

The book will explore in depth the historical achievements of American small business as a primary source of major innovations.

Book Description

The process of technological innovation as it relates to small businesses will be analyzed and its characteristics determined. In *The Culture of Technology*, A. Pacey wrote, "Many of the most significant new ideas in technology have come from small firms and even from individuals working on their own. . .The lone inventor will often need the resources of a large firm to turn invention into marketable innovation, but the key point is that his initial creativity worked best outside bureaucratic limits."

Detailed accounts will be included on many of the exciting, timely, and inspiring technological breakthroughs accomplished by small businesses. Informative, anecdotal, revealing profiles will cover a large number of past and present small firms involved in these breakthroughs.

Important Feature: A key aspect of the book will be the identification and probing examination of common elements found in small firms responsible for historically significant innovations. The objective is to give *other* small businesses practical insights and strategies they can apply to be more effective in achieving the technological innovations that are increasingly critical for competitive survival and business growth.

Large Potential Audience Nationwide

The colorful business success stories used throughout will make the book a stimulating and instructive pleasure to read for scientists, engineers, researchers, managers, small business people, science and business students. The book will combine the often amazing and even perilous struggles of starting and profitably operating a business with the thrill of the perpetual hunt for elusive scientific and technological truth. This will give it considerable appeal to a wide general public as well as professionals.

The thousands of readers that made T.J. Peters and R.H. Waterman's *In Search of Excellence* a best-seller will respond with comparable fervor to this book. The authoritative and intensive focus on individual small businesses and the factors that make them outstanding in the area of technological innovation will assure both popularity and permanent value.

Original Approach: This book will make an appropriate addition to your well-known series of books on business and modern technology, and it offers a completely original approach that has not previously been taken. The assessment of innovation as a small business specialty has not been done on this scale before through case studies, and the analysis of common factors to guide others is entirely new.

In the 20th century, science and business have become our great frontiers for adventure, exploration, and daring, replacing traditional geographic and sea frontiers that summoned buccaneering entrepreneurs during the earlier age of global exploration. The book in its treatment of these frontiers will read like a high-tech thriller, one based entirely on true persons, events, and major R&D conquests. Effective promotion of these attributes through advertising and author talk-show appearances will produce large sales for both hardcover and softcover editions.

Author Qualifications

A.S. Piranti's other books and articles, academic and business affiliations are listed on the accompanying support documents.

Manuscript Delivery Commitment

The research for *Technological Innovation and Small Business in America* is substantially completed, and much of the writing has been done as well. The finished manuscript on computer disk can be available six months following your acceptance of this proposal. I look forward to hearing from you and receiving your approval to submit the complete manuscript for consideration.

The Proposal Package Submitted to the Publisher

Supplementing the above description, the author should also enclose a tentative Table of Contents and sample chapters, if available, from the work in progress. In a letter of enclosure, or as part of the proposal, he might question possible contractual arrangements and make himself available to the publisher for detailed discussion of those terms and others pertaining to the project.

If the proposal is submitted by someone without established credentials, a positive response from the publisher probably at most would be an agreement to examine carefully the finished manuscript. If the proposal comes from someone with a distinguished track record known to the publisher, a contract agreement might be reached for the book prior to its completion if the proposal succeeds in persuasively making the point clear that the book is appropriate for the publisher and that its intended contents give it strong commercial and/or scholarly potential.

CHAPTER 16

CONCLUSION: THE IDEA AND THE PROPOSAL

"A man really writes for an audience of about ten persons. Of course, if others like it, that is clear gain. But if those ten are satisfied, he is content."

Alfred North Whitehead

"Writing is just work—there's no secret. If you dictate or use a pen or type or write with your toes—it is still just work."

Sinclair Lewis

The Idea

We began this proposal-preparing journey with emphasis on using proposals in obtaining grants or other funds to support, develop, and drive ideas that lead to new technologies, products, programs, and progress.

Ideas and proposals go together as naturally and inevitably as bread and cheese, Laurel and Hardy, love and kisses, truth and beauty, income and taxes, Bogart and Bacall, thirst and water, sun and sky. Every fine idea comes eventually to good old Proposal Point. Proposal Point is that inviting natural harbor at land's end of Idea Peninsula or on the biggest island in Concept Archipelago. You have to reach Proposal Point harbor; or those undeveloped ideas and concepts stay land-locked, hidden, and woe of woes, unfunded.

An idea that doesn't eventually require some sort of proposal is more of a notion than an idea and travels on gossamer wings in the opposite direction from reality. Put it down as a cosmic axiom: Every idea worth its salt reaches the proposal stage. Getting it there is part of your challenge, and you should be working on that as well as proposals.

185

The engineer and thinker Lancelot Law Whyte wrote, "There are few experiences quite so satisfactory as getting a good idea. You've had a problem, you've thought about it till you were tired, forgotten it and perhaps slept on it, and then flash! When you weren't thinking about it suddenly the answer has come to you."

An almost equally satisfactory experience, Whyte may have known, is first getting that good idea and then seeing a proposal about the idea succeed in influencing supporters and winning a grant.

The renowned advertising executive Alex Osborn, of Batten, Barton, Durstine & Osborn, in 1941 wrote a famous little book called *How To Think Up* that was widely used in World War II to help people unleash their ingenuity and come up with new ideas for the war effort. Osborn challenged the idea that ideas are a dime a dozen. "Ideas are not dimes but diamonds," he insisted.

That was the period in which big companies began giving big rewards to workers who stepped forward with time and cost-saving ideas. At General Motors, President C.E. Wilson offered a $1,000 Bond for each good idea from an employee, and in five months the company received 31,777 ideas. Osborn quoted the President of B.F. Goodrich, John Collyer, who said, "Nearly all of us have more imagination than we ever put to work. Too often we either do not try hard enough to think things up, or we are too modest to hand in ideas which occur to us. . .Let us put our imaginations on overtime! Let us try to double our volume of suggestions."

Osborn emphasized that no royal road exists to creativity and that the key ingredients for successfully "thinking up" worthwhile ideas are "conscious effort and exercise." Good thinking, the same as push-ups and juggling, benefits from practice. The mind's magic muscles get stronger, bolder, and brighter when you deliberately and conscientiously use them in a regular and persistent creative effort. "The more you exercise your imagination, the more you'll enjoy the effort," promised Osborn.

Osborn's guidelines for creativity emphasize choosing a single target, then turning the imagination loose on it and writing down all the ideas that appear from the tame to the wild. In the cool light of later, the idea seeker reviews these ideas critically and chooses the best for further analytical reflection. After brooding about the ideas, Osborn advises relaxing with other activities, allowing the ideas to age in the mind, take root and shape. Finally, confer with others about the ideas to determine whether they stand up when shared and questioned. "If you still fail to hit your mark, retrace your previous shots, reload your imagination, and fire, and fire again," insisted Osborn, guaranteeing that eventually you will hit the target.

In 1942, when the Allies were often militarily on the run, British statesman Sir Stafford Cripps said, "Lack of imagination is what the United States and Great Britain are suffering from today." New ideas became the leading currency of the time; and acting on the new ideas, such as radar, brought peace. Osborn ended his influential little plea for widespread imaginative thinking with this speculation: "Perhaps in the mind of some modest man there are the makings of the model for the millennium. Will his buried ideas be buried with him, or will he mine them with the dynamite of conscious effort?"

If that modest man or his immodest cousins should develop world-class ideas today, the ideas would certainly be welcome. The world and the human race still desperately need fresh fruits from unshackled imaginations and ideas that change the world and make life better for us all. Yet today as in 1942, the ideas will have to be introduced, explained, and supported with convincing proposals.

Let's say you have applied Osborn's tips or followed your own secret path through the heart of the unknown and dreamed up a really great idea. What next?

The Proposal

What next is up to you. An idea leads to a proposal the way hopeful trails lead to a mountain stream. So the time has come to start writing or to begin assembling the materials to start writing a proposal. As you work on the proposal, consider that small audience of ten people identified by mathematician Alfred North Whitehead as the special group every writer seeks to satisfy.

Just as speakers find it helps to address one person or a few in the crowds before them, writing to a small group should be easier than writing to a vague and threatening throng. At least in the case of the group, you can think specifically about what you need to say to satisfy those concerned.

In the case of a proposal, the audience you wish to satisfy may be even fewer than ten people. Perhaps only one person has to be convinced. But if you can persuade that tiny audience of one or a few individuals with your proposal, you're on your way to the winner's circle.

The challenge with your proposal is to make clear that you know what you're doing and what you intend to do and that you should be helped to do it. Then you may enjoy a payday that would impress practically every writer who ever lived from Homer to Hemingway, from Moses to Mailer.

If these guidelines have prepared you to start, they've achieved their purpose. Guidelines, tips, advice, counsel, suggestions, outlines, and other aids for proposal preparation do nothing more than bring you to the starting point — and provide support when you need it at other points along the way. But the supports can't prepare a proposal. That's your job. Might as well face it.

Don't dwell on the dimension of the task ahead too much, or you could start feeling like Mark Twain's Huck Finn, who declared, "If I'd a knowed what a trouble it was to make a book I wouldn't a tackled it." Readers in that book's first century and now in its second have rejoiced that Huck didn't know about the trouble in advance. When the proposal is finished, you'll rejoice too. So think positively, avoid brooding about the scope of the challenge, and remember that you'll be thrilled, delighted, proud, and funded when the proposals you write start raising R&D money and idea-supporting income.

Getting started, frankly, is always the hardest part of any writing job, proposals not excluded. John Burroughs, the naturalist, warned, "It makes your head hot and your feet cold, and it stops the digesting of your food." Forget about John Burroughs' cold feet and other distractions. "Get it down. Take chances," urged author William Faulkner. Even if it starts slowly or badly, "it's the only way you can do anything really good," Faulkner declared. Great counsel! There's a proposal to write, and the only sure way to start is to start.

Once the process has really begun in earnest, work momentum often takes over, enthusiasm becomes your collaborator, and you find yourself carried along at an accelerating pace to the finish line. Soon there you are with six finished copies of an excellent, vigorous, lucid, technically persuasive proposal ready to submit for approvals and a grant.

A final reminder. . .when your proposal wins the grant, contract, or support you seek, then you have the project to carry out just as you described it in the proposal. Do the project to the best of your ability, but keep in mind that a project generally isn't a career. Don't you have more ideas waiting for proposals?

Since every good idea deserves and demands a good proposal, keep your proposal-writing hat, tools, and skills handy. While the current project makes progress toward its scheduled completion date, why not start preparing your next proposal for another good idea.

REFERENCES

Alsop, Joseph and Stewart Alsop. *The Reporter's Trade* (New York: Reynal & Co., 1958).

Baida, Peter. "Management Babble," *American Heritage*, April 1985.

Barrass, Robert. *Scientists Must Write* (London: Chapman and Hall, 1978).

Barzun, Jacques. "Calamophobia or Hints Toward a Writer's Discipline," *The Writers Book*, edited by Helen Hull (New York: Harper & Brothers, 1950).

Beveridge, W.I.B. *Seeds of Discovery* (New York: W.W. Norton, 1980).

Beveridge, W.I.B. *The Art of Scientific Investigation* (New York: W.W. Norton, 1957).

Bly, Robert W. and Gary Blake. *Technical Writing* (New York: McGraw-Hill, 1982).

Bogosian, Eric. *Sex, Drugs, Rock & Roll* (New York: HarperCollins Publishers, 1991).

Brande, Dorothea. *Becoming a Writer* (New York: Harcourt, Brace and Co., 1934).

Buchwald, Art. "The Great Data Famine," *Washington Post*, September 28, 1969.

Coleman, Peter and Ken Brambleby. *The Technologist as Writer* (New York: McGraw-Hill, 1969).

Cook, Desmond. *Program Evaluation and Review Technique: Applications in Education* (Washington, D.C.: Superintendent of Documents, 1971).

189

Einstein, Albert. *Ideas and Opinions* (New York: Crown Publishers, 1954).

Evans, Bergen. *The Spoor of Spooks* (New York: Alfred A. Knopf, 1954).

Famous Writers Course, Principles of Good Writing, Volumes I and II (Westport, Connecticut: Famous Writers School, 1960).

Faulkner, William, quoted in *A Writer Teaches Writing* by Donald M. Murray (Boston: Houghton Mifflin, 1968).

Flesch, Rudolf. *The Art of Readable Writing* (New York: Harper & Brothers, 1949).

Follett, Wilson. *Modern American Usage* (New York: Grosset & Dunlap, Inc., 1970).

Foundation Center. *Conducting Evaluations: Three Perspectives* (New York: The Foundation Center, 1980).

Foundation Directory, 1991 Edition (New York: The Foundation Center, 1991).

Golde, Roger A. and Jules J. Schwartz. "Be sure you say enough when you write your R&D proposals," *Industrial Research and Development*, March 1983.

Greiner, Larry E. and Robert O. Metzger. *Consulting to Management* (Englewood Cliffs, N.J.: Prentice-Hall, 1983).

Gunning, Robert. *The Technique of Clear Writing* (New York: McGraw-Hill, 1968).

Hammarskjöld, Dag. *Markings* (New York: Alfred A. Knopf, 1964).

Holtz, Herman. *How to Succeed as an Independent Consultant* (New York: John Wiley & Sons, 1983).

Hook, J.N. *Hook's Guide to Good Writing* (New York: Ronald Press Co., 1962).

Hoover, Hardy. *Essentials for the Scientific and Technical Writer* (New York: Dover, 1980).

InKnowVation Newsletter, Innovation Development Institute, Swampscott, Massachusetts, February 1989.

Inman, B.R. and Daniel F. Burton. "Technology and Competitiveness," *Scientific American*, January 1991.

Inventor-Entrepreneur Network Newsletter, Zimmer Foundation, Ann Arbor, Michigan, December 1990.

Jacquette, F. Lee and Barbara I. Jacquette. "What Makes a Good Proposal," *Foundation News*, January/February 1973 (Reprinted by The Foundation Center, New York).

Jewkes, John, David Sawers, Richard Stillerman. *The Sources of Invention* (London: The Macmillan Company, 1958).

Kelley, Robert E. *Consulting, The Complete Guide to a Profitable Career*, Revised Edition (New York: Charles Scribner's Sons, 1986).

King, Lester S. and Charles G. Roland. *Scientific Writing* (American Medical Association, Division of Scientific Publications, 1968).

Kolin, Philip C. *Successful Writing At Work*, Third Edition (Lexington, Massachusetts: D.C. Heath & Company, 1990).

Lenzner, Robert. *The Great Getty* (New York: Crown Publishers, 1985).

Lesko, Matthew. *The Computer Data and Database Source Book* (New York: Avon, 1984).

Lewis, Sinclair, quoted in *A Writer Teaches Writing* by Donald M. Murray (Boston: Houghton Mifflin, 1968).

Lubkin, Yale Jay. "Getting the Contract," *Aerospace & Defense Science*, Volume 9, No. 6, June 1990.

Mack, Karin and Eric Skjei. *Overcoming Writing Blocks* (Los Angeles, California: J.P. Tarcher, 1979).

Mandell, Steven L. *Computers and Data Processing* (St. Paul, Minnesota: West Publishing Co., 1979).

Mayer, Robert A. "What will a foundation look for when you submit a grant proposal?" *Library Journal*, July 1972 (Reprinted by The Foundation Center, New York).

Michigan Conference on R&D Opportunities Summary (Ann Arbor, Michigan: MERRA, April 1989).

Michigan High Tech 90 Conference Summary (Ann Arbor, Michigan: MERRA, September 1990).

Mitchell, Richard. *Less Than Words Can Say* (Boston: Little, Brown and Company, 1979).

Mullins, Carolyn J. *The Complete Writing Guide* (Englewood Cliffs, N.J.: Prentice-Hall, 1980).

National Science Foundation. *Grants for Scientific and Engineering Research* (Washington, D.C.: National Science Foundation, NSF 83–57).

National Science Foundation. *Small Business Guide to Federal R&D Funding Opportunities* (Washington, D.C.: National Science Foundation, Office of Small Business R&D, 1983).

Nelson, J. Raleigh. *Writing the Technical Report* (New York: McGraw-Hill, 1940).

New Science, "Computers should be a plus, not a pain," Office of the Vice President for Research and the Office of Marketing Communications, Wayne State University, Volume 5, 1991.

Nielsen, Waldemar A. *The Big Foundations* (New York: Columbia University Press, 1972).

Osborn, Alex. *How To Think Up* (New York: McGraw-Hill, 1942).

Pacey, A. *The Culture of Technology* (Cambridge, Massachusetts: MIT Press, 1983).

Paris, Barry. "Unconquerable," Profile of Marcia Davenport, *The New Yorker*, April 22, 1991.

Pearsall, Thomas E. and Donald H. Cunningham. *How to Write for the World of Work* (New York: Holt, Rinehart & Winston, 1978).

Peters, T.J. and R.H. Waterman, Jr. *In Search of Excellence* (New York: Warner Books, 1984).

Profiles of Leading Proprietary Technology-Based, Small R&D Businesses (Ann Arbor, Michigan: MERRA, January 1991).

Sherman, Theodore A. *Modern Technical Writing* (Englewood Cliffs, N.J.: Prentice-Hall, 1966).

Smith, Frank R. "Engineering Proposals," *Handbook of Technical Writing Practices*, Volume 1, edited by Stello Jordon, Joseph M. Kleinman, and H. Lee Shimberg (New York: Wiley-Interscience, 1971).

Strunk, William Jr. and E.B. White. *The Elements of Style* (New York: The Macmillan Company, 1959).

Thomas, Lewis. *Late Night Thoughts on Listening to Mahler's Ninth Symphony* (New York: The Viking Press, 1983).

Ueland, Brenda. *If You Want to Write*, Second Edition (St. Paul, Minnesota: Graywolf Press, 1987).

VanOrden, Naola. "Critical-Thinking Writing Assignments in General Chemistry," *The Journal of Chemical Education*, The University of Texas, Austin, Texas, Volume 64, No. 6, June 1987.

Wallich, Paul and Elizabeth Corcoran. "The Analytical Economist," *Scientific American*, January 1991.

Weiss, Edmond H. *The Writing System for Scientists and Engineers* (Englewood Cliffs, N.J.: Prentice-Hall, 1981).

White, Virginia P. *Grants* (New York: Plenum Press, 1975).

Whyte, Lancelot Law. "Where Do Those Bright Ideas Come From," *Harper's*, July 1951.

Williams, Cecil B. and Allan H. Stevenson. *A Research Manual* (New York: Harper & Brothers, 1951).

Young, Jordan R. *How To Become a Successful Freelance Writer*, Third Edition (Anaheim, California: Moonstone Press, 1983).

Zinsser, William. *Writing to Learn* (New York: Harper & Row, 1988).

[Numerous grant program solicitations and requests for proposals (RFPs) were consulted, and many supplied informative materials and advice on proposal preparation. Such publications are useful resources on the subject. They also serve as current catalogs of opportunities. Solicitations and RFPs issue notices about projects seeking proposals for you to investigate and consider. Thus it pays to examine grant program solicitations and RFPs regularly. They are invitations to prepare and submit your proposals with rewards ahead when you succeed.]

INDEX